Jan Heerda

The exponent of Hölder calmness for polynomial systems

Jan Heerda

The exponent of Hölder calmness for polynomial systems

Connections between polynomial degree and exponent of Hölder calmness

Südwestdeutscher Verlag für Hochschulschriften

Impressum / Imprint

Bibliografische Information der Deutschen Nationalbibliothek: Die Deutsche Nationalbibliothek verzeichnet diese Publikation in der Deutschen Nationalbibliografie; detaillierte bibliografische Daten sind im Internet über http://dnb.d-nb.de abrufbar.

Alle in diesem Buch genannten Marken und Produktnamen unterliegen warenzeichen-, marken- oder patentrechtlichem Schutz bzw. sind Warenzeichen oder eingetragene Warenzeichen der jeweiligen Inhaber. Die Wiedergabe von Marken, Produktnamen, Gebrauchsnamen, Handelsnamen, Warenbezeichnungen u.s.w. in diesem Werk berechtigt auch ohne besondere Kennzeichnung nicht zu der Annahme, dass solche Namen im Sinne der Warenzeichen- und Markenschutzgesetzgebung als frei zu betrachten wären und daher von jedermann benutzt werden dürften.

Bibliographic information published by the Deutsche Nationalbibliothek: The Deutsche Nationalbibliothek lists this publication in the Deutsche Nationalbibliografie; detailed bibliographic data are available in the Internet at http://dnb.d-nb.de.

Any brand names and product names mentioned in this book are subject to trademark, brand or patent protection and are trademarks or registered trademarks of their respective holders. The use of brand names, product names, common names, trade names, product descriptions etc. even without a particular marking in this works is in no way to be construed to mean that such names may be regarded as unrestricted in respect of trademark and brand protection legislation and could thus be used by anyone.

Coverbild / Cover image: www.ingimage.com

Verlag / Publisher:
Südwestdeutscher Verlag für Hochschulschriften
ist ein Imprint der / is a trademark of
AV Akademikerverlag GmbH & Co. KG
Heinrich-Böcking-Str. 6-8, 66121 Saarbrücken, Deutschland / Germany
Email: info@svh-verlag.de

Herstellung: siehe letzte Seite /
Printed at: see last page
ISBN: 978-3-8381-3442-0

Zugl. / Approved by: Berlin, Humboldt-Universität zu Berlin, Diss., 2005

Copyright © 2012 AV Akademikerverlag GmbH & Co. KG
Alle Rechte vorbehalten. / All rights reserved. Saarbrücken 2012

to G.

Acknowledgments

I would first and foremost like to thank my supervisor Bernd Kummer for his persistent support. He was the person whose knowledge and understanding allowed me to rapidly immerse myself in what was a new subject to me when I began researching. Without his hints and ideas this work would never have been possible.

Also, I thank Diethard Klatte for helpful dialogues and the opportunity to revise my thoughts at the ›Optimization and Applications‹ seminar 2010 in Zurich, as well as my colleagues at the Humboldt University of Berlin for several fruitful discussions.

And of course it should not go without mentioning my gratefulness to my family and friends for all their love and backup and for giving me the strength to continue striving forwards.

Zusammenfassung

Diese Arbeit befasst sich mit Untersuchung der Hölder Calmness, eines Stabilitätskonzeptes, das man als Verallgemeinerung des Begriffs der Calmness erhält. Ausgehend von Charakterisierungen dieser Eigenschaft für Niveaumengen von Funktionen, werden, unter der Voraussetzung der Hölder Calmness, Prozeduren zur Bestimmung von Elementen dieser Mengen analysiert. Ebenso werden hinreichende Bedingungen für Hölder Calmness studiert.

Da Hölder Calmness (nichtleerer) Lösungsmengen endlicher Ungleichungssysteme mittels (lokaler) Fehlerabschätzungen beschrieben werden kann, werden auch Erweiterungen der lokalen zu globalen Ergebnissen diskutiert.

Als Anwendung betrachten wir speziell den Fall von Niveaumengen von Polynomen bzw. allgemeine Lösungsmengen polynomialer Gleichungen und Ungleichungen. Eine konkrete Frage, die wir beantworten wollen, ist die nach dem Zusammenhang zwischen dem größten Grad der beteiligten Polynome sowie dem Typ, d.h. dem auftretenden Exponenten, der Hölder Calmness des entsprechenden Systems.

Schlagworte: Hölder Calmness, Stabilität, Fehlerabschätzung, Polynomiale Ungleichungssysteme, Hörmander-Łojasiewicz-Ungleichung.

Abstract

This thesis is concerned with an analysis of Hölder calmness, a stability property derived from the concept of calmness. On the basis of its characterization for (sub)level sets, we will cogitate about procedures to determine points in such sets under a Hölder calmness assumption. Also sufficient conditions for Hölder calmness of (sub)level sets and of inequality systems will be given and examined.

Further, since Hölder calmness of (nonempty) solution sets of finite inequality systems may be described in terms of (local) error bounds, we will as well amplify the local propositions to global ones.

As an application we investigate the case of (sub)level sets of polynomials and of general solution sets of polynomial equations and inequalities. A concrete question we want to answer here is, in which way the maximal degree of the involved polynomials is connected to the exponent of Hölder calmness or of the error bound for the system in question.

Key words: Hölder calmness, stability, error bounds, polynomial inequality systems, Hörmander-Łojasiewicz inequality.

Contents

1 Introduction   1

2 Notation and definitions   7
    2.1 Basic notation 7
    2.2 Hölder calmness 8
    2.3 Polynomials 11

3 Error bounds for systems of inequalities and equalities   13
    3.1 Known general results 13
    3.2 Hunting q 23
         3.2.1 The one-dimensional case 23
         3.2.2 More than one dimension 26
    3.3 The Tarski-Seidenberg principle 29

4 Hölder calmness – conditions and characterizations   39
    4.1 The basic theorem 43
    4.2 Iteration schemes for Hölder calmness 50
    4.3 Applying the algorithms 70
         4.3.1 Arbitrary initial points 70
         4.3.2 Application to disturbed optimization problems 75
    4.4 Assigned linear inequality systems 82
    4.5 Sufficient conditions 91

Contents

5 Polynomials   **107**
  5.1 Level sets of polynomials 107
  5.2 Polynomial systems 123

Bibliography   **135**

1 Introduction

An interesting topic regarding optimization programming is the question of (at least local) stability of solutions. This is due to the fact that the parameters and initial values may be not known exactly or cannot be determined accurate, in particular in the case of multilevel programming where one uses preliminary results to solve a problem. So, what one wants to have, is that perturbations of the parameters in the program will only cause predictable changes on the solution. Or, said with other words, that the magnitude of perturbation gives a bound on the change of the solutions of a perturbed problem.

There are several concepts of stability – one of them is *calmness*, a kind of weak local upper Lipschitz property (cf. [KK02b] for a deeper insight). Various Lipschitzian properties in terms of several generalized derivatives have been studied in [RW98] and [KK06].

Now, since we are talking about calmness as a Lipschitz type property, a manifest idea is to generalize this notion to Hölder characteristics, which was also recently done by Kummer [Kum09]. Earlier investigations on this topic have been done by Alt [Alt83], Klatte [Kla85], [Kla94] and Gfrerer [Gfr87] and derived sufficient conditions for Hölder type stability with exponent $q = \frac{1}{2}$.

A subject closely related to the above topic of stability are *error bounds*. Here one is interested in measuring the distance to whatever solution set using easy to calculate residual functions related to the given problem. Hence we want to have some proposition stating that points almost meeting the given conditions are also close to a solution. The motivation to study this arises from

1 Introduction

contemplating about termination criteria in computer implementations of iterative algorithms. (A summary of the theory and application of error bounds may be found in the survey paper of Pang [Pan97].)

Usually the sets in question are described in terms of inequalities and equations – and also many mathematical optimization programs can be treated as such (for instance think about linear programming problems in primal-dual form, nonlinear complementarity problems or general nonlinear optimization programs for differentiable functions via KKT optimality conditions). So we consider systems of type

$$g(x) \leq 0, \ h(x) = 0$$
$$\text{where } g = (g_1, \ldots, g_m), \ h = (h_1, \ldots, h_{m_h}) : X \to \mathbb{R},$$

with nonempty solution set S. An obvious residual function then is

$$r(x) = \|\max\{0, g(x)\}\| + \|h(x)\|,$$

and we want to have some conclusion like

$$\forall\, x \in K \subset X : \operatorname{dist}(x, S) \leq L r(x)^q,$$

where $L > 0$ and $q \in (0, 1]$ are constants.

The well known paper of Hoffman [Hof52] was the first work on this field and he showed that for (finite) systems of affine functions on $X = \mathbb{R}^n$ the above error bound statement holds with $K = \mathbb{R}^n$ and $q = 1$. This result yields just (global) calmness for $S : \mathbb{R}^k \to \mathbb{R}^n$ with $S(p_1, p_2) := \{\, x \in \mathbb{R}^n \mid g(x) \leq p_1,\, h(x) = p_2 \,\}$ at any point $(0, \bar{x}) \in \operatorname{gph} S$.

But in general the exponent q is less than 1 (if it exists at all) – even for (sub)level sets of monomials in one dimension – so here at most Hölder calmness is possible. Using the Hörmander-Łojasiewicz inequality Luo/Luo [LL94] and Luo/Pang [LP94] proved Hölder calmness for systems of polynomials and also analytic functions. But since the Hörmander-Łojasiewicz inequality is based on the Tarski-Seidenberg principle one only knows that there is an exponent for Hölder calmness but cannot specify it exactly this way.

This work is structured as follows:

In Chapter 2 we introduce the necessary notion and give the basic definitions. Chapter 3 contains an overview of several results regarding error bounds and a brief inspection of the Hörmander-Łojasiewicz inequality, including a crash course to semialgebraic sets and the Tarski-Seidenberg principle (see Section 3.3). The main purpose of this chapter is to review this (mathematically elegant) approach to general Hölder calmness in view of its usage to find concrete exponents. Unfortunately we will see that it is not possible to get an satisfying explicit magnitude this way (cf. Proposition 18 and the subsequent comment).

The first main part of the thesis is Chapter 4 where we analyze conditions and characterizations of Hölder calmness. On the basis of a characterization of (local) Hölder calmness for (sub)level sets given by Kummer [Kum09], we present a global characterization (Theorem 25), thus augmenting results of Wu and Ye [WY02a] as well as Ng and Zheng [NZ00] (Section 4.1).

Using this characterizations we then cogitate about procedures to determine points in (sub)level sets under the calmness $[q]$ assumption and also without this premise (Sections 4.2 and 4.3). Particular interesting results are the characterization of Hölder calmness via some relative slack condition (Theorem 30) and that one may characterize Hölder calmness via linear convergence of appropriate algorithms (cf. in particular Lemma 32 and Algorithm 4). Since the considered algorithms need starting points nearby, we also analyzed what would happen for arbitrary starting points (Theorem 33). Moreover we apply the theory to necessary opti-

1 Introduction

mality conditions, gaining that for KKT points calmness $[q]$ does not depend on the description of the KKT set (Lemma 34) and that – under additional conditions – the aforementioned algorithm computes Fritz-John points (Lemma 35).

Afterwards we will shortly contemplate about crucial index sets for Hölder calmness of inequality systems (Section 4.4), prior to specifying and examining sufficient conditions for Hölder calmness of (sub)level sets and of systems of inequations (Section 4.5). The obtained results generalize prior statements of Wu and Ye [WY02a] for convex systems to the non-convex case also (Theorems 40 and 41). Based on these findings we then get a sufficient condition for calmness $[1/d]$ of d-times continuously differentiable functions (Lemma 45, offering a sufficient condition for the general descending requirement (4.71)). This is the result which than bridges to Chapter 5. As a specific conclusion before giving way to the next chapter we will get that for C^2 functions with regular Hessian it holds calmness $[\frac{1}{2}]$ (Corollary 46).

In Chapter 5 we apply our findings to the case of (sub)level sets of polynomials and to general solution sets of polynomial equations and inequalities. First we get that quadratic polynomials have at least calm $[\frac{1}{2}]$ level sets (Theorem 49). Then, based on propositions achieved by Ng and Zheng [NZ00], we generalize this to a global error bound for (sub)level sets of quadratic polynomials with exponent $\frac{1}{2}$, where we especially go into detail with respect to the constant L, which may be computed (Theorem 50).

Subsequently we use this result to show that over compact sets there is an quadratic error bound for systems of one quadratic and finitely many affine functions (Theorem 56). In contrast to prior statements of Luo, Pang and Ralph [LPR96] we do not need any nonnegativity condition here, and unlike to an analogous result of Luo and Sturm [LS00] we have a direct proof, which also allows to figure out the constant L. Insofar those propositions are put into bigger framework. On the way we alternatively prove known results of Luo and Sturm [LS00] and of Luo, Pang and Ralph [LPR96] (Corollaries 54 and 55).

Unfortunately one cannot gain such general statement as mentioned before for (sub)level sets of polynomials of degree 4 and greater or systems containing more than one quadratic function (see Examples 12 and 16). And we did not find any result for (sub)level sets of cubic polynomials.

2 Notation and definitions

2.1 Basic notation

Let X and P be Banach spaces. We write $\|\cdot\|$ for the norm, $d(\cdot,\cdot)$ for the (induced) distance and $\operatorname{dist}(y,M) := \inf\{\, d(y,\tilde{y}) \mid \tilde{y} \in M \,\}$ for the distance between a point y and a subset M of the respect space, where $\operatorname{dist}(y,\emptyset) := +\infty$. Further let

$$\mathbb{B}(x,\varepsilon) := \{\, y \mid d(y,x) \leq \varepsilon \,\}$$

denote the the closed ball with radius ε around x, and for subsets M we put

$$\mathbb{B}(M,\varepsilon) := \bigcup_{x \in M} \mathbb{B}(x,\varepsilon),$$

which is M with some additional ε-neighbourhood. In any case we may use some subscript to indicate the underlying space or norm – in particular for subspaces Y of X it is $\mathbb{B}_Y$ the closed ball in Y w.r.t. the induced distance d_Y. Moreover $\|\cdot\|_2$ will denote the Euclidean and $\|\cdot\|_\infty$ the maximum norm in finite dimension.

Writing $F : X \rightrightarrows P$, we mean that F is a multifunction between the two spaces, i.e. $F(x) \subseteq P$ for $x \in X$ and we denominate by $\operatorname{dom} F := \{\, x \in X \mid F(x) \neq \emptyset \,\}$ the domain of F and by $\operatorname{gph} F := \{\, (x,p) \in X \times P \mid p \in F(x) \,\}$ its graph.

Another notation used at times in this paper is $o(t)$, which denotes a quantity of type $\lim_{t \to 0^+} \frac{o(t)}{t} = 0$.

In addition, for real numbers a and vectors $x = (x_1, \ldots, x_n) \in \mathbb{R}^n$, let $a^+ = \max\{0,a\}$ and $x^+ = (x_1^+, \ldots, x_n^+)$ respectively.

2 Notation and definitions

Several derivative notions

As usual we write $\partial_i g$ for the i-th partial derivative of a function $g : \mathbb{R}^n \to \mathbb{R}$, Dg for its Fréchet derivative or the Jacobian, and $D^2 g$ for its second Fréchet derivative or the Hessian. Clarke's generalized Jacobian is denoted by $\partial^{Cl} g$. Also we will use the lower subderivative[1]

$$d^- g(x)(u) := \liminf_{\substack{t \to 0^+ \\ u' \to u}} \frac{g(x + tu') - g(x)}{t},$$

and the (one-sided) directional derivative

$$g'_i(x; u) := \lim_{t \to 0^+} \frac{g_i(x + tu) - g_i(x)}{t}.$$

2.2 Hölder calmness

Stability analysis is interested in the stability of local solutions $x \in X$ of generalized equations $p \in F(x)$ for given canonical parameters $p \in P$, where $F : X \rightrightarrows P$ is a closed multifunction, i.e. $\operatorname{gph} F$ is closed[2]. Or to put it into other words the problem setting is:

> For a given closed multifunction $F : X \rightrightarrows P$ and parameter $p \in P$ find some $x \in X$ such that $p \in F(x)$. (2.1)

For functions $f : X \to P$ one identifies $f(x)$ and $F(x) = \{f(x)\}$. Then F is closed in particular for continuous f and one is directly in the setting in this case.

---

[1] Here we follow the notation of [RW98]. For a reflection about the several names see [RW98, p. 345].

[2] In particular this definition yields that F is closed iff F^{-1} is closed.

2.2 Hölder calmness

To analyze stability one studies the behaviour of the solution sets to (2.1)

$$S(p) := F^{-1}(p) = \{\, x \in X \mid p \in F(x) \,\}$$

near some particular solution.

Remark 1. If S is a solution set for some inequality system $g(x) \leq 0$ and $h(x) = 0$ where $g = (g_1, \ldots, g_m)$, $h = (h_1, \ldots, h_{m_h})$ are functions with values in $\mathbb{R}$, then S is closed if g is lower semi-continuous and h continuous.

An overview to several notions of stability can be found in the book of Klatte and Kummer [KK02b]. In this work we will only consider a stability property called calmness.

Following [KK02b] we say for $S : P \rightrightarrows X$ and $(\bar{p}, \bar{x}) \in \operatorname{gph} S$:

Definition 1. S is *calm* at $(\bar{p}, \bar{x})$, iff exist ε, δ, $L > 0$ such that for all $p \in \mathbb{B}(\bar{p}, \delta)$ holds

$$S(p) \cap \mathbb{B}(\bar{x}, \varepsilon) \subset \mathbb{B}(S(\bar{p}), L\, \|p - \bar{p}\|), \text{ i.e.}$$

$$\forall x \in S(p) \cap \mathbb{B}(\bar{x}, \varepsilon) : \ \operatorname{dist}(x, S(\bar{p})) \leq L \, \|p - \bar{p}\|. \qquad (2.2)$$

The constant L is called the *rank* of calmness.

Note. Let us denote here by the way that calmness is the weakest of all Lipschitz type stability conditions which is also a constraint qualification, i.e. guarantees that the existence of KKT points is necessary for solutions of optimization problems with standard constraints – which is (beside its relation to error bounds) another important motivation to study calmness in the first place.

We may generalize this definition to a Hölder type characteristic in the familiar way (cf. [Kum09] as well):

2 Notation and definitions

Definition 2. S is called *Hölder calm with exponent q* or *calm $[q]$* at $(\bar{p}, \bar{x})$, iff exist $q \in (0, 1]$, ε, δ, $L > 0$ such that for all $p \in \mathbb{B}(\bar{p}, \delta)$ holds
$$S(p) \cap \mathbb{B}(\bar{x}, \varepsilon) \subset \mathbb{B}(S(\bar{p}), L \, \|p - \bar{p}\|^q), \text{ i.e.}$$

$$\forall x \in S(p) \cap \mathbb{B}(\bar{x}, \varepsilon) : \ \text{dist}(x, S(\bar{p})) \leq L \, \|p - \bar{p}\|^q. \tag{2.3}$$

Remark 2. Obviously, if a multifunction S is calm $[q]$ at $(\bar{p}, \bar{x}) \in$ gph S then S is also calm $[q']$ at $(\bar{p}, \bar{x}) \in$ gph S for every $0 < q' \leq q \leq 1$.

Remark 3. Definition 2 is equivalent to the existence of some $q \in (0, 1]$, ε, δ, $L > 0$ such that for all $p \in \mathbb{B}(\bar{p}, \delta)$ holds
$$x \in S(p) \cap \mathbb{B}(\bar{x}, \varepsilon) \to S(\bar{p}) \cap \mathbb{B}(x, L \, \|p - \bar{p}\|^q) \neq \emptyset,$$
i.e. $\forall x \in S(p) \cap \mathbb{B}(\bar{x}, \varepsilon) \, \exists x' \in S(\bar{p}) : \ \|x - x'\| \leq L \, \|p - \bar{p}\|^q$.

Remark 4. If $\bar{x} \in \text{int } S(\bar{p})$ then S is trivially calm $[q]$ (for any q) at $(\bar{p}, \bar{x})$, since then one may just choose $\varepsilon > 0$ s.t. $\mathbb{B}(\bar{x}, \varepsilon) \subset S(\bar{p})$ and hence $\forall \, x \in \mathbb{B}(\bar{x}, \varepsilon) : \ \text{dist}(x, S(\bar{p})) = 0$.

Thus in the following we will always consider $\bar{x} \in S(\bar{p}) \backslash \text{int } S(\bar{p})$ without saying so explicitly.

Remark 5. One easily sees that S is not calm $[q]$ at $(\bar{p}, \bar{x}) \in$ gph S if and only if
$$\exists \{(p^k, x^k)\} \subset \text{gph } S :$$
$$\left(\forall \, k \in \mathbb{N} : x^k \neq \bar{x} \wedge p^k \neq \bar{p} \right) \wedge \left((p^k, x^k) \to (\bar{p}, \bar{x}) \right)$$
$$\wedge \left(\forall \, \xi^k \in \underset{\xi \in S(\bar{p})}{\text{argmin}} \|x^k - \xi\| : \ \frac{\|p^k - \bar{p}\|}{\|x^k - \xi^k\|^{1/q}} \to 0 \right).$$

Remark 6. Another trivial statement (following directly from the definition) is that a multifunction S is calm $[q]$ at some point

$(\bar p, \bar x) \in \mathrm{gph}\, S$ if

$$\exists\, \lambda, \varepsilon, \delta > 0 \ \forall\, p \in \mathbb{B}(\bar p, \delta)\setminus\{\bar p\} \ \forall\, x \in S(p) \cap \mathbb{B}(\bar x, \varepsilon):$$
$$\lambda \|x - \bar x\| \leq \|p - \bar p\|^q. \tag{2.4}$$

2.3 Polynomials

Notation. We call $\alpha = (\alpha_1, \ldots, \alpha_n) \in \mathbb{N}_0^n$ a *multiindex*, the *length* of α is $|\alpha| = \sum_{i=1}^n \alpha_i$ and we define $M_n^d := \{\, \alpha \in \mathbb{N}_0^n \mid |\alpha| = d \,\}$ and $M_n^{\leq d} := \cup_{i=0}^d M_n^i$ to be the set of all multiindices with length d and length at most d, respectively.

Note. The set M_n^d has $\binom{n+d-1}{d} = \frac{(n+d-1)!}{d!(n-1)!}$ elements.

Notation. A real *monomial* on $\mathbb{R}^n$ is any function $h: \mathbb{R}^n \to \mathbb{R}$ of the form

$$h(x) = a \cdot x^\alpha := a \cdot \prod_{i=1}^n x_i^{\alpha_i} \qquad (a \in \mathbb{R}\setminus\{0\}).$$

The length of α then is called the *total degree* of the monomial.

Now $h: \mathbb{R}^n \to \mathbb{R}$ is called a *polynomial* if it is a sum of monomials, i.e. if there are (different) multiindices $\alpha^1, \ldots \alpha^l \in \mathbb{N}_0^n$ such that

$$h(x) = \sum_{i=1}^l a_{\alpha^i} x^{\alpha^i} \qquad (a_{\alpha^i} \in \mathbb{R}\setminus\{0\}).$$

The *degree of a polynomial* is the largest total degree d of the involved monomials. If we put $a_\alpha = a_{\alpha^i}$ for $\alpha = \alpha^i$ and $a_\alpha = 0$ else, then we may write h as

$$h(x) = \sum_{\alpha \in M_n^{\leq d}} a_\alpha x^\alpha \quad \text{or shortly} \quad h(x) = \sum_{|\alpha| \leq d} a_\alpha x^\alpha.$$

Finally, we call a polynomial h *homogeneous* if its monomials with nonzero coefficients all have the same total degree, i.e. if it has the

form
$$h(x) = \sum_{\alpha \in M_n^d} a_\alpha x^\alpha = \sum_{|\alpha|=d} a_\alpha x^\alpha$$
for some $d \in \mathbb{N}_0$.

Note. If h is a homogeneous polynomial of degree d than h is a homogeneous function of degree d, i.e. $h(rx) = r^d h(x)$ for all $x \in \mathbb{R}^n$ and $r \in \mathbb{R}$.

3 Error bounds for systems of inequalities and equalities

3.1 Known general results

Consider the following inequality system in finite dimension

$$g_i(x) \leq 0, \ i = 1, \ldots, m, \quad \text{and} \quad h_j(x) = 0, \ j = 1, \ldots, m_h, \qquad (3.1)$$

for functions $g_i, h_j : \mathbb{R}^n \to \mathbb{R}$ and let S be the solution set of this system, which is assumed to be nonempty. We will denote the vector functions $g = (g_1, \ldots, g_m)$ and $h = (h_1, \ldots, h_{m_h})$.

For stability analysis the question arises whether or not and under which conditions an error bound of (3.1) in terms of some residual function exists, that is:

Are there positive constants L, q and a subset T of $\mathbb{R}^n$ such that with the residual $r(x)$, which is a non-negative valued vector function fulfilling »$r(x) = 0$ if and only if $x \in S$«, holds

$$\forall x \in T : \ \text{dist}(x, S) \leq L \|r(x)\|^q \, ? \qquad (3.2)$$

A natural and popular choice for the residual function is

$$r(x) := (g_1(x)^+, \ldots, g_m(x)^+, |h_1(x)|, \ldots, |h_{m_h}(x)|)$$

which leads to $\|r(x)\| = \|(g(x)^+, h(x))\|$ or (in finite dimension) equivalently to $\|g(x)^+\| + \|h(x)\|$.

The first publication dealing with error bounds is the paper of Hoffman [Hof52]. He showed

Proposition 1 (Hoffman's error bound). *If g and h are affine linear functions, i.e. $g(x) = Ax + a$ and $h(x) = Bx + b$ for some matrices A, B and vectors a, b of appropriate dimensions[1], then there exists some constant $L > 0$ depending on A and B only such that (for arbitrary norm)*

$$\forall x \in \mathbb{R}^n : \ dist(x, S) \leq L(\|(Ax + a)^+\| + \|Bx + b\|). \qquad (3.3)$$

But such simple bound does not hold even for (general) polynomial mappings:

Example 1. Let $g : \mathbb{R} \to \mathbb{R}$ be given by $g(x) = x^d$, with $1 < d \in \mathbb{N}$, and $S := \{\, x \mid g(x) \leq 0 \,\}$. Then clearly $S = \{0\}$ if d is even and $S = \mathbb{R}_0^-$ if d is odd. So, given any $x \in \mathbb{R}$, we have

$$dist(x, S) = \begin{cases} |x|, & \text{if } d \text{ is even} \\ 0, & \text{if } d \text{ is odd and } x \leq 0 \\ |x|, & \text{if } d \text{ is odd and } x > 0 \end{cases}$$

and

$$|g(x)^+| = \begin{cases} |x|^d, & \text{if } d \text{ is even} \\ 0, & \text{if } d \text{ is odd and } x \leq 0 \\ |x|^d, & \text{if } d \text{ is odd and } x > 0 \end{cases}$$

So for all $x \in \mathbb{R}$ holds $dist(x, S) \leq |g(x)^+|^{1/d}$, but there is no $L > 0$ such that $dist(x, S) \leq L|g(x)^+|$ for all x near the origin:

---

[1]In fact Hoffman originally considered only systems of linear inequalities, but as every system of equalities may be written as two systems of inequalities one gets the given result.

3.1 Known general results

Assuming this, it would follow that there is some $\varepsilon > 0$ such that in particular $|x| \leq L|x|^d$ for all $0 < x < \varepsilon$, i.e. $L \geq \frac{1}{|x|^{d-1}} \xrightarrow[x \to 0]{} \infty$, which is a contradiction.

Moreover the following example due to Luo and Pang (see [LP94, Example 4.3] and [LPR96, 2.3.14 Example]) shows that in general the error bound is only local:

Example 2. *For the solution set*

$$S := \{ (x_1, x_2) \mid x_1 x_2 = 0, -x_1 \leq 0, -x_2 + 1 \leq 0 \}$$
$$= \{ (0, x_2) \mid x_2 \geq 1 \}$$

holds:

$$\mathrm{dist}((t,0), S) = t \text{ for } t \geq 1 \text{ and maximum-norm,}$$

but

$$\|(x_1 x_2, -x_1, -x_2 + 1)(t, 0)^+\|_\infty = 1.$$

Nevertheless there are general propositions regarding error bounds of solution sets of polynomial systems.

For the case of a single real polynomial $h : \mathbb{R}^n \to \mathbb{R}$ Hörmander [Hör58, Lemma 1 and 2] proved

Proposition 2 (Hörmander's error bound). *There are positive constants L, q and a (possibly negative) constant q' such that (for the Euclidean norm $\|\cdot\|_2$)*

$$\forall x \in \mathbb{R}^n : \quad \mathrm{dist}_2(x, S) \leq L(1 + \|x\|_2^2)^{q'} |h(x)|^q. \qquad (3.4)$$

This was extended by Luo and Luo [LL94, Theorem 2.2] to sets S given by systems of polynomial equalities and inequalities. Considering the new polynomial $f : \mathbb{R}^{n+m} \to \mathbb{R}$ given by

$$f(x, z) = \sum_{i=1}^{m} (g_i(x) + z_i^2)^2 + \sum_{j=1}^{m_h} h_j(x)^2 \qquad (3.5)$$

3 Error bounds for systems of inequalities and equalities

and using that $x \in S$ iff $f(x,z) = 0$ for $z_i = z_i(x) = \sqrt{(-g_i(x))^+}$, they obtained – applying Hörmander's result[2]

Proposition 3 (Luo/Luo error bound). *There exist constants $L > 0$, $q > 0$ and $q' \geq 0$ such that*

$$\forall x \in \mathbb{R}^n : \ dist_2(x,S) \leq L(1+\|x\|_2)^{q'}(\|g(x)^+\|_2 + \|h(x)\|_2)^q. \quad (3.6)$$

Remark 7. Obviously results (3.4) and (3.6) can be written in terms of arbitrary compact subsets, i.e. for any compact K exist positive L and q such that

$$\forall x \in K : \ dist_2(x,S) \leq L|h(x)|^q,$$

and

$$\forall x \in K : \ dist_2(x,S) \leq L(\|g(x)^+\|_2 + \|h(x)\|_2)^q,$$

respectively.

Remark 8. Independently of Hörmander, Łojasiewicz [Łoj59, 17. Théorème p. 124] – previously announced without proof in [Łoj58] – gave a bound similar to (3.4) even for a single analytic function, stating that for each compact subset K of $\mathbb{R}^n$ exist $L > 0$ and $q > 0$ such that

$$\forall x \in K : \ dist_2(x,S) \leq L|h(x)|^q. \quad (3.7)$$

Using construction (3.5) Luo and Pang [LP94, Theorem 2.2] generalized this to sets defined by analytic equations and inequations, concluding that for every compact $K \subset \mathbb{R}^n$ exist positive constants L, q such that

$$\forall x \in K : \ dist_2(x,S) \leq L(\|g(x)^+\|_2 + \|h(x)\|_2)^q. \quad (3.8)$$

---

[2]Actually [LL94, Theorem 2.1] – which is the presentation of Hörmander's proposition in the paper of Luo and Luo – does not mention the square of $\|x\|_2$ in (3.4), which is then also 'lost' in (3.6). But this is of course not a problem since $1 + \|x\|^2 \leq (1 + \|x\|)^2$.

3.1 Known general results

Remark 9. Although (3.4), (3.6), (3.7) and (3.8) were proved for the Euclidean norm only[3], by equivalence of norms in finite dimension, these propositions certainly hold for arbitrary norms.

We have the following connection between error bounds and calmness $[q]$:

Lemma 4. *Let* $S(p_1, p_2) := \{ x \in \mathbb{R}^n \mid g(x) \leq p_1 \wedge h(x) = p_2 \}$ *with* $g = (g_1, \ldots, g_m)$ *lower semicontinuous and* $h = (h_1, \ldots, h_{m_h})$ *continuous functions into* $\mathbb{R}$.

Then, with $0 < q \leq 1$, *(the closed multifunction)* S *is calm* $[q]$ *at* $(0, \bar{x}) \in \mathrm{gph}\, S$ *if and only if*

$$\exists \varepsilon > 0, L > 0 \,\forall x \in \mathbb{B}(\bar{x}, \varepsilon):$$
$$\mathrm{dist}(x, S(0)) \leq L \left(\|g(x)^+\| + \|h(x)\| \right)^q.$$

Proof.

($\Leftarrow$) Take any $p = (p_1, p_2) \in \mathbb{R}^{m+m_h}$ and $x \in S(p) \cap \mathbb{B}(\bar{x}, \varepsilon)$. If $p_{1i} \leq 0$ we then have $g_i(x)^+ = 0$ and else $0 \leq g_i(x)^+ \leq p_{1i}$, so in any case $\|g(x)^+\| \leq \|p_1\|$ and further $h(x) = p_2$, i.e. $\|h(x)\| = \|p_2\|$.
Hence by assumption

$$\mathrm{dist}(x, S(0)) \leq L \left(\|p_1\| + \|p_2\| \right)^q \leq 2^q L \|p\|^q.$$

($\Rightarrow$) Since S is calm $[q]$ at $(0, \bar{x})$ we find $\varepsilon, \delta, L > 0$ s.t.

$$\forall p \in \mathbb{B}(0, \delta) \,\forall x \in S(p) \cap \mathbb{B}(\bar{x}, \varepsilon) : \mathrm{dist}(x, S(0)) \leq L \|p\|^q.$$

Put $\bar{\varepsilon} := \min\{\varepsilon, \delta, 1\}$ and let $x \in \mathbb{B}(\bar{x}, \bar{\varepsilon})$.
If $\| (g(x)^+, h(x)) \| \leq \bar{\varepsilon} \leq \delta$ we thus have

$$\mathrm{dist}(x, S(0)) \leq L \|(g(x)^+, h(x))\|^q \leq L_1 \left(\|g(x)^+\| + \|h(x)\| \right)^q,$$

---
[3]This is mainly because the proof of the Hörmander-Łojasiewicz inequality for semialgebraic sets is based on elimination of quantifiers over real closed fields and one may write (in-)equalities regarding $\|\cdot\|_2$ easily as polynomial (in-)equalities (using $\|\cdot\|_2^2$).

3 Error bounds for systems of inequalities and equalities

for some $L_1 > 0$ independent of x.[4]

Else it holds

$$\text{dist}(x, S(0)) \leq \|x - \bar{x}\| \leq \bar{\varepsilon} \leq \|(g(x)^+, h(x))\|.$$

And since in the case $\|(g(x)^+, h(x))\| \geq 1$ it is $\|(g(x)^+, h(x))\|^q \geq 1$ as well and else $\|(g(x)^+, h(x))\| \leq \|(g(x)^+, h(x))\|^q$, this yields

$$\text{dist}(x, S(0)) \leq L_2 \left(\|g(x)^+\| + \|h(x)\|\right)^q.$$

Thus we conclude that for all $x \in \mathbb{B}(\bar{x}, \bar{\varepsilon})$:

$$\text{dist}(x, S(0)) \leq \max\{L_1, L_2\} \left(\|g(x)^+\| + \|h(x)\|\right)^q. \quad \square$$

Remark 10. Note that, for closed multifunctions S given as a solution set of inequalities and equations, to characterize calmness $[q]$ at $(0, \bar{x}) \in \text{gph } S$ one doesn't need to consider only points with small function values (as the definition of Hölder calmness would suggest).

Also, if we abandon the demand for total closedness of S and conform with closedness of the set $S(0)$ then we won't need continuous functions h_i but lower semicontinuous $|h_i|$ (because $h_i(x) = 0$ iff $|h_i(x)| \leq 0$).

A direct corollary of Lemma 4 for (sub)level sets is

Corollary 5. *Let $g : X \to \mathbb{R}$ be lower semicontinuous with $g(\bar{x}) = 0$ for some $\bar{x} \in X$.*

Then its sublevel set map $S(p) = \{x \mid g(x) \leq p\}$ is calm $[q]$ at $(0, \bar{x})$ if and only if

$$\exists \varepsilon > 0 \; \exists L > 0 \; \forall x \in \mathbb{B}(\bar{x}, \varepsilon) : dist(x, S(0)) \leq L(g(x)^+)^q; \quad (3.9)$$

---

[4]For $\|\cdot\| = \|\cdot\|_1$ we would have $L_1 = L$ and thus the inequality follows from the equivalence of norms in $\mathbb{R}^k$.

3.1 Known general results

or equivalently (since S is closed)[5]

$$\exists \varepsilon > 0 \; \exists L > 0 \; \forall x \in \mathbb{B}(\bar{x}, \varepsilon) \setminus S(0) \; \exists x' \in S(0):$$
$$\|x - x'\| \leq L g(x)^q. \qquad (3.10)$$

And calmness [q] of level set maps $S(p) = \{\, x \mid h(x) = p \,\}$ for continuous functions $h : X \to \mathbb{R}$ is equivalent to

$$\exists \varepsilon > 0 \; \exists L > 0 \; \forall x \in \mathbb{B}(\bar{x}, \varepsilon) : \operatorname{dist}(x, S(0)) \leq L|h(x)|^q. \qquad (3.11)$$

Remark 11. Note that there is a strong connection between Hölder calmness of the level set $S_=(p) = \{\, x \mid h(x) = p \,\}$ and the sublevel set $S_\leq(p) = \{\, x \mid h(x) \leq p \,\}$ for continuous functions h:

Obviously it is $S_= \subset S_\leq$. Hence, if $S_=(p)$ is calm [q] at $(0, \bar{x})$, then we have

$$\operatorname{dist}(x, S_\leq(0)) \leq \operatorname{dist}(x, S_=(0)) \leq L|h(x)|^q = L(h(x)^+)^q$$

for all $x \notin S_\leq(0)$ near $\bar{x}$, i.e. we have also calmness [q] of $S_\leq(p)$ at $(0, \bar{x})$.

Using the result of Luo and Luo we finally get

Corollary 6. *For every*

$$S(p_1, p_2) := \{\, x \in \mathbb{R}^n \mid g(x) \leq p_1 \wedge h(x) = p_2 \,\},$$

where $g = (g_1, \ldots, g_m)$ *and* $h = (h_1, \ldots, h_{m_h})$ *are real polynomials on* $\mathbb{R}^n$, *exists some* $q > 0$ *such that* S *is calm* [q] *at* $(0, \bar{x})$ *for every* $\bar{x} \in S(0)$.

Proof. If $S(0)$ is empty this is a tautology. So let's suppose $S(0) \neq \emptyset$. Then by (3.6) there exist $\tilde{L}, q, q' > 0$ such that for every $x \in \mathbb{R}^n$

---
[5]Note that now only $x \notin S(0)$ near $\bar{x}$ are considered and thus in particular $g(x) > 0$ is already ensured.

3 Error bounds for systems of inequalities and equalities

holds

$$\mathrm{dist}(x, S(0)) \leq \tilde{L}(1 + \|x\|)^{q'}(\|g(x)^+\| + \|h(x)\|)^q.$$

Now take any $\bar{x} \in S(0)$, $\varepsilon > 0$ and $(p_1, p_2) \in \mathbb{R}^m \times \mathbb{R}^{m_h}$, $x \in S(p_1, p_2) \cap \mathbb{B}(\bar{x}, \varepsilon)$. Since $\|x\| \leq \varepsilon + \|\bar{x}\|$ and $\|p_1\| + \|p_2\| \leq 2\|(p_1, p_2)\|$ we get

$$\begin{aligned}\mathrm{dist}(x, S(0)) &\leq \tilde{L}(1 + \varepsilon + \|\bar{x}\|)^{q'}(\|p_1\| + \|p_2\|)^q \\ &\leq L\|(p_1, p_2)\|^q\end{aligned} \quad (3.12)$$

where $L = \tilde{L}(1 + \varepsilon + \|\bar{x}\|)^{q'} 2^q$. □

Remark 12. Of course, using the Łojasiewicz result and [LL94, Theorem 2.2], Corollary 6 is equally true for analytic functions.

But note that such result does not hold in general for (sub)level sets of general C^∞ functions:

Example 3. *Consider the non-analytic C^∞ function*

$$g(x) = \begin{cases} e^{-(1/x^2)}, & x \neq 0 \\ 0, & x = 0. \end{cases}$$

Since $S(0) := \{\, x \in \mathbb{R} \mid g(x) \leq 0 \,\} = \{0\}$ it is $\mathrm{dist}(x, S(0)) = |x|$ for all $x \in \mathbb{R}$. So due to $\lim_{x \to 0} \frac{|x|^r}{e^{-(1/x^2)}} = \lim_{y \to \infty} \frac{e^{(y^2)}}{y^r} = \infty$ for each $r > 0$, $S(p)$ cannot be Hölder calm at $(0, 0)$.

The above Lemma 4 (and its corollary 5) states that calmness [q] is a local error bound property with exponent q.

If one assumes a system of convex functions on $\mathbb{R}^n$ then it holds even a global equivalence (the next Lemma and its proof are an adaption of [Li97, Theorem 3.3]):

3.1 Known general results

Lemma 7. *Let $g_i : \mathbb{R}^n \to \mathbb{R}$, $i = 1, \ldots, m$, be convex functions. Then the multifunction S defined as*

$$S(p) := \{\, x \in \mathbb{R}^n \mid \bigwedge_{i=1}^n (g_i(x) \le p_i) \,\},$$

and with $S(0) \ne \emptyset$, is calm $[q]$ at every point $(0, \bar{x}) \in \operatorname{gph} S$ if and only if for any compact set $K \subset \mathbb{R}^n$ exists some constant $L_K > 0$ such that

$$\forall\, x \in K : \operatorname{dist}(x, S(0)) \le L_K \|g(x)^+\|^q. \qquad (3.13)$$

Proof. The backward direction is clearly true, so we consider calmness $[q]$ at every $(0, \bar{x}) \in \operatorname{gph} S$, i.e. (by Lemma 4) for each $\bar{x} \in S(0)$ there are $\varepsilon_{\bar{x}}, L_{\bar{x}} > 0$ s.t.

$$\forall\, x \in \mathbb{B}(\bar{x}, \varepsilon_{\bar{x}}) : \operatorname{dist}(x, S(0)) \le L_{\bar{x}} \|g(x)^+\|^q. \qquad (3.14)$$

Now fix $x^* \in S(0)$ and $y^* \in K$ (w.l.o.g. $K \ne \emptyset$ because then (3.13) is trivially true). Set $r := \sup\{\, \|y - y^*\| \mid y \in K \,\}$ and define

$$S_r := \{\, x \in S(0) \mid \|x\| \le 2(r + \|y^*\|) + \|x^*\| \,\},$$

which is a compact set.

Obviously $S_r \subset \bigcup_{\bar{x} \in S_r} \mathbb{B}^o(\bar{x}, \varepsilon_{\bar{x}})$ and so by compactness there are points $\bar{x}_1, \ldots, \bar{x}_k \in S_r$ fulfilling

$$S_r \subset \bigcup_{j=1}^k \mathbb{B}^o(\bar{x}_j, \varepsilon_{\bar{x}_j}). \qquad (3.15)$$

Now for $x \in \mathbb{R}^n$ let x' denote a nearest element in $S(0)$, i.e. $x' \in S(0)$ and $\|x - x'\| = \operatorname{dist}(x, S(0))$. Then for any $x \in K$ we have

$$\|x'\| \le \|x\| + \|x - x'\| \le \|x - y^*\| + \|y^*\| + \|x - x^*\|$$
$$\le r + \|y^*\| + \|x\| + \|x^*\| \le 2(r + \|y^*\|) + \|x^*\|,$$

i.e. $x' \in S_r$ and hence by (3.15) it is $x' \in \mathbb{B}^o(\bar{x}_j, \varepsilon_{\bar{x}_j})$ for some $j = 1, \ldots, k$. Since $\mathbb{B}^o(\bar{x}_j, \varepsilon_{\bar{x}_j})$ is an open set, there exists $\theta \in (0, 1)$ such that
$$x_\theta := \theta x + (1 - \theta)x' \in \mathbb{B}^o(\bar{x}_j, \varepsilon_{\bar{x}_j}).$$
By (3.14) we obtain
$$\operatorname{dist}(x_\theta, S(0)) \leq L_{\bar{x}_j} \|g(x_\theta)^+\|^q.$$
Because of convexity of the g_i and since $g_i(x') \leq 0$ for all i, we get
$$g_i(x_\theta) \leq \theta g_i(x) + (1 - \theta)g_i(x') \leq \theta g_i(x),$$
and hence $\|g(x_\theta)^+\| \leq \theta \|g(x)^+\|$.

Moreover we have by definition of x_θ and with $x_\theta' \in S(0)$ selected to be a nearest point to x_θ in $S(0)$
$$\|x - x'\| \leq \|x - x_\theta'\| \leq \|x - x_\theta\| + \|x_\theta - x'\|$$
$$= (1 - \theta)\|x - x'\| + \|x_\theta - x_\theta'\|,$$
which implies
$$\theta \operatorname{dist}(x, S(0)) \leq \operatorname{dist}(x_\theta, S(0)).$$
So, for $L_K := \theta^{q-1} \max_{j=1,\ldots,k} L_{\bar{x}_j}$, it follows
$$\operatorname{dist}(x, S(0)) \leq \theta^{-1} \operatorname{dist}(x_\theta, S(0))$$
$$\leq \theta^{-1} \max_{j=1,\ldots,k} L_{\bar{x}_j} \|g(x_\theta)^+\|^q$$
$$\leq L_K \|g(x)^+\|^q. \qquad \square$$

Remark 13. The only part we need convexity in the above proof is for the estimate $\|g(x_\theta)^+\| \leq \theta \|g(x)^+\|$. So, if we get this in a different way, Lemma 7 holds for general systems of continuous functions on $\mathbb{R}^n$.

3.2 Hunting q

An interesting question is whether there is a connection between the maximal degree of the polynomials defining S as in the corollary and the exponent of Hölder calmness – and what this relation looks like.

By (3.3) we have calmness for affine functions (i.e. both the maximal degree and the exponent are 1) and Example 1 may indicate that a similar relation holds even for greater degrees of the polynomials involved, meaning that if the maximal degree is d then one has Hölder calmness with exponent $1/d$.

In the first subsection we will analyze the one-dimensional case, where we will easily get the result as aforementioned. In the case of more than one dimension we will then show that the same result may be attained for some very special cases, but also that it is not possible to generalize the Hoffman proof for linear functions to general polynomials – not even homogeneous ones. (As will be demonstrated in the subsequent Chapter 5 this is not by chance.)

3.2.1 The one-dimensional case

Let $h(x) := \sum_{i=0}^{d} a_i x^i$ be a one-dimensional polynomial with degree $d \in \mathbb{N}$, i.e. $a_d \neq 0$. In particular thus h is not constant.

Lemma 8. *For every root $\bar{x}$ of h there is some $c > 0$ such that for all x near $\bar{x}$*
$$|h(x)| \geq c \cdot |x - \bar{x}|^d.$$

Proof. The inequality holds trivially for $x = \bar{x}$, so let $x \neq \bar{x}$. It is clear that $h(x) = (x - \bar{x})^k \cdot p(x)$ for some $1 \leq k \leq d$ and a real polynomial p with $\deg p = d - k$ and $p(\bar{x}) \neq 0$.

For $k = d$ thus $p(x)$ is nonzero but constant, which yields for all $x \in \mathbb{R}$ that $|h(x)| = c\,|x - \bar{x}|^d$ with $c = |p(x)|$.

3 Error bounds for systems of inequalities and equalities

Now let $k < d$. Then

$$\frac{|h(x)|}{|x-\bar{x}|^d} = \frac{|x-\bar{x}|^k}{|x-\bar{x}|^d}|p(x)| = \frac{1}{|x-\bar{x}|^{d-k}}|p(x)| \xrightarrow[x\to\bar{x}]{} \infty;$$

so there is some $\varepsilon > 0$ such that $|h(x)| \geq |x-\bar{x}|^d$ for all $x \in [\bar{x}-\varepsilon, \bar{x}+\varepsilon]$. □

Corollary 9. *Let h be a real polynomial of degree $d > 0$ and S the set of its real roots. Further let K be a compact subset of $\mathbb{R}$ such that $S \cap K \neq \emptyset$.*

Under these conditions there is some $L > 0$ such that

$$\forall\, x \in K : \mathrm{dist}(x, S \cap K) \leq L\,|h(x)|^{\frac{1}{d}}.$$

Proof. Let $S \cap K = \{\bar{x}_1, \ldots, \bar{x}_k\}$ with $\bar{x}_i < \bar{x}_{i+1}$, $i = 1, \ldots, k-1$.

First assume $K = [a, b]$. We separate this closed interval into the closed subintervals $K_1 := [a, \frac{\bar{x}_2 + \bar{x}_1}{2}]$, $K_i := [\frac{\bar{x}_{i-1} + \bar{x}_i}{2}, \frac{\bar{x}_i + \bar{x}_{i+1}}{2}]$, $i = 2, \ldots, k-1$, and $K_k := [\frac{\bar{x}_{k-1} + \bar{x}_k}{2}, b]$. By Lemma 8, for each $\bar{x}_i$, there are ε_i and $c_i > 0$ such that

$$\forall\, x \in (\bar{x}_i - \varepsilon_i, \bar{x}_i + \varepsilon_i) : c_i\,|x - \bar{x}_i|^d \leq |h(x)|.$$

Now the sets $E_i := K_i \setminus (\bar{x}_i - \varepsilon_i, \bar{x}_i + \varepsilon_i)$ are compact and thus $m_i := \min_{x \in E_i} \frac{|h(x)|}{|x-\bar{x}_i|^d}$ exists. Moreover $m_i > 0$, because $h(x) \neq 0$ on E_i by construction. Here we put $m_i = +\infty$ if $E_i = \emptyset$.

With $\lambda_i = \min\{c_i, m_i\} > 0$ it follows

$$\forall\, x \in K_i : \lambda_i\,|x - \bar{x}_i|^d \leq |h(x)|,$$

and, because of $\mathrm{dist}(x, S \cap K) = |x - \bar{x}_i|$ for all $x \in K_i$, we have, setting $L := \max_{i=1,\ldots,k} \lambda_i^{-1}$, that

$$\forall\, x \in K : \mathrm{dist}(x, S \cap K) \leq L\,|h(x)|^{\frac{1}{d}}.$$

3.2 Hunting q

In the general case $K = \bigcup_{j=1}^{s}[a_j, b_j]$ with $b_j < a_{j+1}, j = 1, \ldots, s-1$, it holds by the above part

$$\forall\, x \in [a_j, b_j] : \operatorname{dist}(x, S \cap [a_j, b_j]) \leq L_j\, |h(x)|^{\frac{1}{d}}$$

for some $L_j > 0$. As $S \cap K \supset S \cap [a_j, b_j]$ this yields

$$\forall\, x \in [a_j, b_j] : \operatorname{dist}(x, S \cap K) \leq L_j\, |h(x)|^{\frac{1}{d}}$$

and thus, for $L = \max_j L_j$, we get

$$\forall\, x \in K : \operatorname{dist}(x, S \cap K) \leq L\, |h(x)|^{\frac{1}{d}}. \qquad \square$$

As a consequence of this 'global' statement we obtain

Corollary 10. *Let h be a real polynomial of degree $d > 0$. Then the level set $S(p) := \{\, x \in \mathbb{R} \mid h(x) = p \,\}$ is Hölder calm with exponent $q = \frac{1}{d}$ at $(0, \bar{x})$ for every $\bar{x} \in S(0)$.*

Note. If $h \equiv 0$ then $S(p) = \emptyset$ for all $p \neq 0$ and we have (proper) calmness. And if $h \equiv c \neq 0$ then $S(0) = \emptyset$, so there is no $\bar{x} \in S(0)$.

Proof. Let $h \not\equiv 0$ and consider any $\bar{x} \in S(0)$ (if $S(0) = \emptyset$ we are already done). As the set $S(0)$ of zeros of h in $\mathbb{R}$ has at most d elements, there is some $\mathbb{B}(\bar{x}, \varepsilon)$ which does not contain any other element of $S(0)$ than $\bar{x}$ itself. By the above proposition thus there is some $L > 0$ such that $L|h(x)|^{1/d} \geq \operatorname{dist}(x, S(0) \cap \mathbb{B}(\bar{x}, \varepsilon)) \geq \operatorname{dist}(x, S(0))$ for all $x \in \mathbb{B}(\bar{x}, \varepsilon)$. $\qquad \square$

***Remark* 14.** Unfortunately we cannot use the proof of Corollary 10 in more than one dimension, because in general we cannot separate the roots of polynomials in higher dimension (just take $h(x) = x_1 x_2$ to see this) and it does not hold a statement similar to Lemma 8 as the following example shows:

Example 4. *Consider $h : \mathbb{R}^2 \to \mathbb{R}$ defined by $h(x) = x_1^2 x_2 - x_2^4$. Then for $x = (\varepsilon^2, \varepsilon)$ with $\varepsilon \in (0, 1)$ it holds:*

$$|h(x)| = |\varepsilon^4 \varepsilon - \varepsilon^4| = \varepsilon^4(1-\varepsilon) < \varepsilon^4 = |-x_2^4|,$$

i.e. it is not true that $|h(x)| = \left|\sum_{|\alpha| \leq d} a_\alpha x^\alpha\right| \geq \left|\sum_{|\alpha|=d} a_\alpha (x - \bar{x})^\alpha\right|$ for all x near $\bar{x} = 0$.

Another statement about estimates regarding roots of real polynomials in one variable is the following proposition, which (together with its proof) is cited from the monograph of Coste [Cos00] (cf. [Cos02] as well):

Proposition 11 ([Cos02, Proposition 1.3]). *Take h as above to be a real polynomial in one variable of degree d. Then for every root $z \in \mathbb{C}$ of h one has the estimate*

$$|z| \leq \max_{i=0,\ldots,d-1} \left(d \frac{|a_i|}{|a_d|}\right)^{1/(d-i)}.$$

Proof. Set $M := \max_{i=0,\ldots,d-1} \left(d \frac{|a_i|}{|a_d|}\right)^{1/(d-i)}$. Then for all $x \in \mathbb{C}$ with $|x| > M$ it is of course $|a_i| < \frac{|a_d|}{d} |x|^{d-i}$ for each $i = 0, \ldots, d-1$. Hence it holds

$$\left|\sum_{i=0}^{d-1} a_i x^i\right| \leq \sum_{i=0}^{d-1} |a_i||x|^i < |a_d||x|^d = |a_d x^d|,$$

and thus $h(x) \neq 0$. □

3.2.2 More than one dimension

Lemma 12. *Let the monomial $h : \mathbb{R}^n \to \mathbb{R}$ be given by $h(x) = a \cdot \prod_{i=1}^n x_i^{\alpha_i}$ with $a \neq 0$ and total degree d greater than zero. Then*

3.2 Hunting q

exists $L > 0$ such that for each $x \in \mathbb{R}^n$ holds

$$\mathrm{dist}(x, S(0)) \leq L|h(x)|^{1/d};$$

so in particular $S(p) := \{\, x \in \mathbb{R}^n \mid h(x) = p \,\}$ is calm $[1/d]$ at $(0, \bar{x})$ for every $\bar{x} \in S(0)$.

Proof. Put $I = \{\, i \mid \alpha_i \neq 0 \,\}$ which is not empty since

$$d = \sum_{i=1}^n \alpha_i > 0.$$

It holds for every $x \in \mathbb{R}^n$

$$|h(x)| = |a| \prod_{i=1}^n |x_i|^{\alpha_i} = |a| \prod_{i \in I} |x_i|^{\alpha_i}$$
$$\geq |a|(\min_{i \in I} |x_i|)^{\sum_{i \in I} \alpha_i} = |a|(\min_{i \in I} |x_i|)^d.$$

Further it is $S(0) = \{\, x \mid \vee_{i \in I} x_i = 0 \,\}$ and thus

$$\mathrm{dist}(x, S(0)) = \min_{i \in I} |x_i|$$

for all $x \in \mathbb{R}^n$.

So we have for all $x \in \mathbb{R}^n$

$$\mathrm{dist}(x, S(0))^d \leq |a|^{-1}|h(x)|;$$

which yields the proposition for $L = |a|^{-\frac{1}{d}}$. $\square$

But what about general polynomials or at least homogeneous ones? Hoffman's proof [Hof52] of (3.3) is based on two lemmas of Agmon [Agm54, Lemma 2.2 and 2.3]. Maybe one could modify them in an appropriate manner? If we adapt [Agm54, Lemma 2.3] to the case of a level set of one homogeneous polynomial, we get the following

3 Error bounds for systems of inequalities and equalities

Lemma 13. *Let $S = \{\, x \in \mathbb{R}^n \mid h(x) := \sum_{|\alpha|=d} a_\alpha x^\alpha = 0 \,\}$ where $d \in \mathbb{N}$, $\alpha \in \mathbb{N}_0^n$. Then exists $c > 0$ such that for all $x \in E = \{\, x \in \mathbb{R}^n \mid x \notin S \wedge dist(x, S) = \|x\| \,\}$ holds*

$$c \cdot dist(x, S)^d \leq |h(x)|. \tag{3.16}$$

Note. Of course $0 \in S$, so $\mathrm{dist}(x, S) = \|x\|$ just means that 0 is the point of S nearest to x.

Proof. Consider

$$E_1 := E \cap \mathrm{bd}\, \mathbb{B}(0,1) = \{\, x \in \mathbb{R}^n \mid dist(x, S) = \|x\| \wedge \|x\| = 1 \,\}.$$

It holds[6]

$$x \in E \iff \frac{x}{\|x\|} \in E_1,$$

so $E_1 \neq \emptyset$ if we assume E to be nonempty. And moreover $E_1 \subset \mathbb{R}^n$ is bounded and closed, i.e. compact.

As $f(x) := |h(x)|$ is continuous thus there exists $\tilde{x} \in E_1$ with $f(\tilde{x}) = \min_{x \in E_1} f(x)$. In particular $\tilde{x} \notin S$, so $c := f(\tilde{x}) > 0$, which yields that $f(x) \geq c > 0$ for all $x \in E_1$. Thus $f(\frac{x}{\|x\|}) \geq c > 0$ for all $x \in E$. And since

$$f\Big(\frac{x}{\|x\|}\Big) = \Big|\sum_{|\alpha|=d} a_\alpha \Big(\frac{x}{\|x\|}\Big)^\alpha\Big| = \frac{1}{\|x\|^d}\Big|\sum_{|\alpha|=d} a_\alpha x^\alpha\Big|,$$

we get (3.16) for every $x \in E$. $\square$

But regrettably one cannot guarantee $E \neq \emptyset$ for degrees d greater than 1, so we do not have an useful proposition here[7] – not mentioning other problems arising when trying to use this lemma for some adaption of Hoffman's proof in the case $d > 1$.

---

[6]It is $x \notin S$ iff $\frac{1}{\|x\|^d} h(x) \neq 0$ iff $h(x/\|x\|) \neq 0$ iff $x/\|x\| \notin S$; and $\mathrm{dist}(x, S) = \|x\|$ iff $\mathrm{dist}(x/\|x\|, S) = 1$ for $x \notin S$.

[7]In the linear case S is a hyperplane, so nonemptiness is clear.

Example 5. *Consider the solution set*

$$S = \{\, x = (x_1, x_2) \in \mathbb{R}^2 \mid x_1 x_2 = 0 \,\}$$
$$= \{\, (x_1, x_2) \in \mathbb{R}^2 \mid x_1 = 0 \vee x_2 = 0 \,\},$$

and let $x = (x_1, x_2) \notin S$.

Since $(0, x_2), (x_1, 0) \in S$ and $\|x - (0, x_2)\| = |x_1|$ as well as $\|x - (x_1, 0)\| = |x_2|$, we get $\mathrm{dist}(x, S) \leq \min_{i=1,2} |x_i|$. But

$$\|x\|_k = \sqrt[k]{|x_1|^k + |x_2|^k} > \min_{i=1,2} |x_i|$$

for each $x \notin S$, $1 \leq k < \infty$.

In the remaining case $\|\cdot\|_\infty$ we take

$$S = \{\, x = (x_1, x_2) \in \mathbb{R}^2 \mid x_1^2 - x_2^2 = 0 \,\}$$
$$= \{\, (x_1, x_2) \in \mathbb{R}^2 \mid x_1 = x_2 \vee x_1 = -x_2 \,\}.$$

Then for all $x = (x_1, x_2) \notin S$ it holds $\|x\|_\infty = \max_i |x_i| > \min_i |x_i|$. Because $y = (\tfrac{1}{2}(x_1 - x_2), -\tfrac{1}{2}(x_1 - x_2)) \in S$, it follows $\mathrm{dist}(x, S) \leq \|x - y\|_\infty \leq \tfrac{1}{2}(|x_1| + |x_2|) < \|x\|_\infty$.

We end this little hunt for the exponent at this point, and will come to more coherent approaches. But we already want to underline here that the statement of Lemma 12 is not true in such general form for even general homogeneous polynomials of degree at least 4 (cf. Example 13).

3.3 The Tarski-Seidenberg principle

As the proof of Corollary 6 (which showed Hölder calmness for systems of polynomials) is in the end based on Proposition 2 (Hörmander's error bound), an analysis of the proof of this result may give an answer to our question concerning the connection between

3 Error bounds for systems of inequalities and equalities

the maximal degree of the polynomial system and the exponent of the regarding Hölder calmness.

The main tool Hörmander used is the so called *Tarski-Seidenberg principle*. This principle was first announced by Tarski without a proof in [Tar31], but the publication of a proof lasted until [Tar48]. Later Seidenberg [Sei54] gave a new approach, which is the base for today's versions of the principle used in algebraic geometry. This versions may also be given under the name *Projection theorem for semialgebraic sets*, *Quantifier elimination over real closed fields* or *Transfer principle*.

In the following we want to analyze this approach in order to see what kind of results we may expect with respect to concrete conclusions about exponents. To do so we reproduce parts of [Cos02] in order to give a short introduction into the theory of real algebraic geometry as far as it affects our topic; for further references see [BPR06], [Cos00], [Cos02], [BCR98], [BR90] and [ABR96].

The statements in this area of mathematics are usually made for arbitrary real closed fields. But as we are not interested in such generalization we will consider here only the standard real closed field – the real numbers $\mathbb{R}$.

Definition 3 (Algebraic and semialgebraic sets and functions). Let $\mathbb{R}[X_1, \ldots, X_n]$ denote as customary the set of real polynomials on $\mathbb{R}^n$. For any finite subset $\mathcal{P}$ of $\mathbb{R}[X_1, \ldots, X_n]$ the set of zeros of $\mathcal{P}$ in $\mathbb{R}^n$ is defined as

$$Z(\mathcal{P}) := \left\{ x \in \mathbb{R}^n \mid \bigwedge_{h \in \mathcal{P}} h(x) = 0 \right\}.$$

Such $Z(\mathcal{P})$ are called the *algebraic sets* of $\mathbb{R}^n$.

Note. Using the common construction $h = h_1^2 + \ldots + h_m^2$ one can write any algebraic set $Z(\{h_1, \ldots, h_m\})$ as the set of zeros of only one polynomial.

3.3 The Tarski-Seidenberg principle

A *basic semialgebraic set* of $\mathbb{R}^n$ is any set of the form

$$\left\{ x \in \mathbb{R}^n \,\middle|\, h(x) = 0 \wedge \bigwedge_{g \in \mathcal{Q}} g(x) < 0 \right\}$$

where $h \in \mathbb{R}[X_1, \ldots, X_n]$ and $\mathcal{Q}$ a finite subset of $\mathbb{R}[X_1, \ldots, X_n]$.
Note. By the above note it is clear that sets of the form

$$\left\{ x \in \mathbb{R}^n \,\middle|\, \bigwedge_{h \in \mathcal{P}} h(x) = 0 \wedge \bigwedge_{g \in \mathcal{Q}} g(x) < 0 \right\},$$

with $\mathcal{P}, \mathcal{Q} \subset \mathbb{R}[X_1, \ldots, X_n]$ finite, are basic semialgebraic sets as well.

Now a subset of $\mathbb{R}^n$ is called *semialgebraic* if it is a finite union of basic semialgebraic sets.

Note. The family of semialgebraic sets is closed under the Boolean operations (complementing, finite unions and finite intersections) and is the smallest such family of subsets of $\mathbb{R}^n$ containing all algebraic sets.

Let $M \subset \mathbb{R}^m$, $N \subset \mathbb{R}^n$ be two semialgebraic sets. A mapping $\varphi : M \to N$ is called a *semialgebraic function* if its graph is a semialgebraic set in $\mathbb{R}^{m+n}$.

To dive (just a little bit) deeper into the below propositions, we also need some precise conception of what is meant by a *first-order formula* of the language of ordered fields with parameters in $\mathbb{R}$ (FO formula for short).

Definition 4 (First-order formulas). *FO formulas* are obtained by the following rules:

1. If $p \in \mathbb{R}[X_1, \ldots, X_n]$, then $p(x) = 0$ and $p(x) < 0$ are FO formulas.

2. If ϕ and ψ are FO formulas, then "ϕ and ψ", "ϕ or ψ" and "not ϕ" (denoted as $\phi \wedge \psi$, $\phi \vee \psi$ and $\neg \phi$, respectively) are FO formulas.

3. If ϕ is a FO formula and x a variable ranging over $\mathbb{R}$, then $\exists x \phi$ and $\forall x \phi$ are FO formulas.

Here the formulas achieved by using only rules 1 and 2 are called *quantifier-free*.

Note. Directly by definition one sees that for semialgebraic sets $A \subset \mathbb{R}^n$ we can write the property $x \in A$ using a quantifier-free formula $\phi(x)$ – namely ϕ represents the (finite) "or"-union of the quantifier-free formulas defining the basis sets needed to describe A. Thus in this case we may interpret $x \in A$ directly as a quantifier-free formula.

Moreover every quantifier-free formula defines a semialgebraic set, which can be seen by bringing the formula into disjunctive normal form.

Together we have that $A \subset \mathbb{R}^n$ is semialgebraic if and only if there is a quantifier-free formula $\phi(x_1, \ldots, x_n)$ such that

$$x \in A \iff \phi(x).$$

Having the above notation the Tarski-Seidenberg principle may be stated in the following form:

Proposition 14 (Tarski-Seidenberg principle; [Cos02, Theorem 2.6]). *For every first-order formula $\phi(x_1, \ldots, x_n)$ – having quantifiers or not – the set $S = \{\, x \in \mathbb{R}^n \mid \phi(x) \,\}$ is semialgebraic.*

This may be reformulated as: *Every FO formula is equivalent to a quantifier-free formula, i.e. it is possible to eliminate the quantifiers.*

Based on Tarski-Seidenberg we are able to prove the Hörmander-Łojasiewicz inequality which gives information concerning the relative growth rate of two arbitrary continuous semialgebraic functions. Beforehand we shall estimate the rate of growth of a semialgebraic function in one variable.

3.3 The Tarski-Seidenberg principle

Proposition 15 ([Cos02, Proposition 2.11]). *Let $\varphi : (a, \infty) \to \mathbb{R}$ be a semialgebraic function[8]. Then exist $b \geq a$ and $N \in \mathbb{N}$ s.t.*

$$\forall\, x \in (b, \infty) : |\varphi(x)| \leq x^N. \tag{3.17}$$

Proof. By assumption $\mathrm{gph}\,\varphi$ is a semialgebraic subset of $\mathbb{R}^2$ and thus for some semialgebraic sets

$$G_i := \{\, (x,y) \in \mathbb{R}^2 \mid h_i(x,y) = 0 \wedge \bigwedge_{j=1}^{k_i} g_{ij}(x,y) < 0 \,\},$$

where h_i and g_{ij} are real polynomials, it is

$$\mathrm{gph}\,\varphi = \bigcup_{i=1}^{s} G_i.$$

Here for each polynomial h_i it is $\deg h_i(x,\cdot) > 0$ for all $x \in \mathrm{dom}\,\varphi$:

Otherwise there would be some index i_0 with $(x_0, y_0) \in G_{i_0}$ and $\deg h_{i_0}(x_0, \cdot) = 0$, i.e. $h_{i_0}(x_0, y) = \mathit{const}$ for all y.

And since the finitely many continuous $g_{i_0 j}$ fulfill $g_{i_0 j}(x_0, y_0) < 0$, there is some $\varepsilon > 0$ s.t. $g_{i_0 j}(x_0, y) < 0$ for every index j and each $y \in (y_0 - \varepsilon, y_0 + \varepsilon)$.

Together we would have $(x_0, y) \in \mathrm{gph}\,\varphi$ for $y \in (y_0 - \varepsilon, y_0 + \varepsilon)$, which contradicts that φ is a function into $\mathbb{R}$.

Now put

$$h(x,y) := \prod_{i=1}^{s} h_i(x,y) = \sum_{j=0}^{d} a_j(x) y^j$$

for some $d > 0$ and $a_d \not\equiv 0$. Because a_d is a polynomial in one variable it has only finitely many roots and we may choose $c \geq a$ big enough such that $a_d(x) \neq 0$ for all $x \in (c, \infty)$.

---

[8]Not necessarily a continuous one.

3 Error bounds for systems of inequalities and equalities

Since $\varphi(x)$ is a root of $h(x,\cdot)$, Proposition 11 yields

$$\forall\, x \in (c, \infty): \; |\varphi(x)| \leq \max_{i=0,\ldots,d-1} \left(d\, \frac{|a_i(x)|}{|a_d(x)|} \right)^{\frac{1}{d-i}}.$$

With $a_i(x) = \sum_{j=0}^{l_i} \alpha_{ij} x^j$ we have $|a_i(x)| \leq \sum_{j=0}^{l_i} |\alpha_{ij}||x|^{l_i}$ for $|x| \geq 1$ and $|a_i(x)| \geq \frac{1}{2}|\alpha_{dl_d}||x|^{l_d}$ whenever $|x|$ is large enough, and so it follows for some $\tilde{c} \geq \max\{1,c\}$ that[9] for each $i = 0, \ldots, d-1$ and all $x \in (\tilde{c}, \infty)$

$$\left(d\, \frac{|a_i(x)|}{|a_d(x)|} \right)^{\frac{1}{d-i}} \leq \left(\frac{2dM}{|\alpha_{dl_d}|} x^m \right)^{\frac{1}{d-i}} \leq \max\left\{ \frac{2dM}{|\alpha_{dl_d}|}, 1 \right\} x^m,$$

where $M := \max_{i=0,\ldots,d-1} \sum_{j=0}^{l_i} |\alpha_{ij}|$ and $m := \max_{i=0,\ldots,d-1} \frac{l_i}{l_d}$.

Taking $b = \max\left\{ \tilde{c}, \max\left\{ \frac{2dM}{|\alpha_{dl_d}|}, 1 \right\} \right\}$ and $m+1 \leq N \in \mathbb{N}$ we thus get (3.17). $\square$

Note. Note that in the above proof the strict inequalities do not play any role concerning the values of b or N, but only the equations describing the particular basic semialgebraic sets composing gph φ are important.

Moreover, by construction of the polynomial h, the degree of its coefficients $a_i(x)$ may be quite large – depending on the number s of necessary basic sets G_i and the degree (with respect to x) of the respective polynomials h_i.

Proposition 16 (Hörmander-Łojasiewicz inequality; [Cos02, Theorem 2.12]). *Let $K \subset \mathbb{R}^n$ be a compact semialgebraic set, and let $f, g: K \to \mathbb{R}$ be continuous semialgebraic functions such that*

$$\forall\, x \in K : (g(x) = 0 \to f(x) = 0).$$

---
[9]Depending on the structure of the $a_i(x)$ there may be of course much better estimates.

3.3 The Tarski-Seidenberg principle

Then there exist an integer $N \in \mathbb{N}$ and a constant $C \geq 0$, such that
$$\forall\, x \in K : |f(x)|^N \leq C|g(x)|.$$

Proof. For $t > 0$, set $F_t := \{\, x \in K \mid t|f(x)| = 1\,\}$. Note that, since F_t is closed in K, this set is compact, and because of
$$F_t = \{\, x \in \mathbb{R}^n \mid x \in K \wedge \left((x, \tfrac{1}{t}) \in \operatorname{gph} f \vee (x, -\tfrac{1}{t}) \in \operatorname{gph} f\right)\,\},$$
it is also semialgebraic.

In the case $F_t \neq \emptyset$, by our precondition, the function g does not vanish on F_t and the continuous (and semialgebraic) map $G(x) := \frac{1}{g(x)}$ has a maximum on F_t, which we denote $\theta(t)$. If $F_t = \emptyset$, we set $\theta(t) = 0$.

The so defined function $\theta : (0, \infty) \to \mathbb{R}$ is semialgebraic because of the Tarski-Seidenberg principle and its graph $\operatorname{gph}\theta$ equals
$$\left\{(t, \alpha) \in \mathbb{R} \times \mathbb{R} \,\Big|\, t > 0 \wedge \left(\alpha = \max_{x \in F_t} G(x) \vee \left(\neg \exists\, x \in F_t \wedge \alpha = 0\right)\right)\right\},$$
where $\alpha = \max_{x \in F_t} G(x)$ stands for $(\forall\, x \in F_t\, \forall\, \alpha' : (x, \alpha') \in \operatorname{gph} G \to \alpha' \leq \alpha) \wedge (\exists\, x \in F_t : (x, \alpha) \in \operatorname{gph} G)$.

By Proposition 15, there exist $b > 0$ and $N \in \mathbb{N}$ such that
$$\forall\, t > b : |\theta(t)| \leq t^N,$$
which is equivalent to
$$\forall\, x \in K : \left(0 < |f(x)| < \frac{1}{b} \to \frac{1}{|g(x)|} \leq \frac{1}{|f(x)|^N}\right).$$

If we set D to be the maximum of the continuous function $\frac{|f(x)|^N}{|g(x)|}$ on the compact set $K_b := \{\, x \in K \mid |f(x)| \geq 1/b\,\}$ (observe that on this set the function g does not vanish because of our precondition) and define $C := \max\{D, 1\}$, we obtain for all $x \in K$

$$|f(x)|^N \le C|g(x)|.\qquad\square$$

Note. Here the compact set K_b is not relevant for the exponent N, it only impacts the constant C. The exponent depends solely on the semialgebraic function θ.

As a corollary of the Hörmander-Łojasiewicz inequality and the fact that $\mathrm{dist}(\cdot, S)$ is semialgebraic for semialgebraic sets S (see below), we get the original Hörmander result (cf. Proposition 2), here stated again in a slightly different way:

Corollary 17. *For a polynomial $g : \mathbb{R}^n \to \mathbb{R}$ with zero set $S = \{\, x \mid g(x) = 0 \,\}$ and a compact semialgebraic set $K \subset \mathbb{R}^n$ there are $L > 0$ and $N \in \mathbb{N}$ such that*

$$\forall\, x \in K : \mathrm{dist}(x, S)^N \le L\,|g(x)|. \qquad (3.18)$$

So what do we learn about the exponent N? Examining the aforementioned proofs we see that the description of $\mathrm{gph}\,\theta \subset \mathbb{R}^2$ by a quantifier-free formula plays the important role. Getting it is usually not so easy, but we may state a general first-order formula (amplifying the construction given in the proof of Proposition 16):

It is $(t, \alpha) \in \mathrm{gph}\,\theta$ if and only if

$$-t < 0 \wedge \Big(\big((\forall x\,\forall\alpha' : x \in F_t \wedge (x, \alpha') \in \mathrm{gph}\,(1/g) \to \alpha' \le \alpha)$$
$$\wedge\,(\exists x : x \in F_t \wedge (x, \alpha) \in \mathrm{gph}\,(1/g))\big)$$
$$\vee\,\big(\forall x : x \notin F_t \wedge \alpha = 0\big)\Big),$$

where $(x, \alpha) \in \mathrm{gph}\,(1/g)$ means the polynomial $\alpha\, g(x) = 1$ and

3.3 The Tarski-Seidenberg principle

$x \in F_t$ stands for $(x, 1/t) \in \operatorname{gph dist}_2(\cdot, S)$, i.e.

$$\Big(\forall y : g(y) = 0 \to 1 \leq t^2 \sum (x_i - y_i)^2\Big) \wedge$$
$$\Big(\forall \beta : \big(\forall y : g(y) = 0 \to \beta^2 \leq \sum (x_i - y_i)^2\big) \to \beta t \leq 1\Big).$$

Now we need to apply elimination of quantifiers to this formula. Doing so the degree and number of involved polynomials in general increases. To our knowledge the best known bound of this increase was given by Basu, Pollack and Roy in [BPR96, Theorem 1.3.1] and states:

Proposition 18. *If d is the maximal degree of the s polynomials in a FO formula with quantifiers, then – using an algorithm of quantifier elimination – the degree of the polynomials in the quantifier free formula is at most d^{E_0} and their quantity does not exceed $s^{E_1} d^{E_2}$ with E_0 depending on the number of bound variables and E_1, E_2 dependant on the total number of variables (bound and free).*

Now, as seen in the proof of Proposition 15, both the number and degree[10] of the polynomials involved in the FO formula given above blow up the exponent N. So even for the solution set of one single linear equation the exponent one gets using the quantifier elimination approach may be large – even though we know that $N = 1$ suffices.

---

[10]Checking the formula we see at a glance that the maximal degree is at least 2 – because of $t^2 \sum (x_i - y_i)^2$ it is even 4.

4 Hölder calmness – conditions and characterizations

In this chapter we will collect several useful propositions regarding calmness $[q]$ of multifunctions $S : P \rightrightarrows X$ and in particular (sub-)level sets – or lower level sets as they were called in [RW98] – of lower semicontinuous functions from X to $\mathbb{R}$. These statements will give us a handy tool for the analyzation of the exponent of Hölder calmness for (sub)level sets of polynomials in Chapter 5, but they are also important results in the context of stability of solution sets of (in)equality systems.

The first characterization is Lemma 2.2 in [Kum09] – a similar statement was already given for $q = 1$ in [KK02a, Lemma 3.2] – and shows that calmness $[q]$ is a monotonicity property with respect to two canonically assigned Lipschitz functions: the distance of x to $S(\bar{p})$

$$\mathrm{dist}(x, S(\bar{p}))$$

and the graph-distance

$$\psi_S(x,p) := \mathrm{dist}((p,x), \mathrm{gph}\, S),$$

defined via $d((p,x),(p',x')) = \max\{d(p,p'), d(x,x')\}$ or some equivalent metric in $P \times X$.

4 Hölder calmness – conditions and characterizations

Lemma 19 ([Kum09, Lemma 2.2]). *Let $0 < q \leq 1$. Then the multifunction S is calm $[q]$ at $(\bar{p}, \bar{x}) \in \operatorname{gph} S$ if and only if*

$$\exists \varepsilon > 0 \ \exists \alpha > 0 \ \forall x \in \mathbb{B}(\bar{x}, \varepsilon) : \ \alpha \operatorname{dist}(x, S(\bar{p})) \leq \psi_S(x, \bar{p})^q. \quad (4.1)$$

Note. In other words, calmness $[q]$ at $(\bar{p}, \bar{x})$ is violated iff

$$0 < \psi_S(x_k, \bar{p})^q = o\bigl(\operatorname{dist}(x_k, S(\bar{p}))\bigr) \\ \text{holds for some sequence } x_k \to \bar{x}, \quad (4.2)$$

where $\psi_S(x_k, \bar{p})^q = o\bigl(\operatorname{dist}(x_k, S(\bar{p}))\bigr)$ means that

$$\frac{\psi_S(x_k, \bar{p})^q}{\operatorname{dist}(x_k, S(\bar{p}))} \xrightarrow[k \to \infty]{} 0.$$

Let's recapitulate the proof:

Proof. Let (4.1) hold true. Given $x \in S(p) \cap \mathbb{B}(\bar{x}, \varepsilon)$, it is

$$\psi_S(x, \bar{p})^q \leq d((\bar{p}, x), (p, x))^q = \|p - \bar{p}\|^q$$

and, in consequence, $\alpha \operatorname{dist}(x, S(\bar{p})) \leq \|p - \bar{p}\|^q$, which yields calmness $[q]$ with rank $L = \frac{1}{\alpha}$.

Conversely, let (4.1) be violated, which means that (4.2) is true. Now, given any positive $\delta_k < \psi_S(x_k, \bar{p})^q$, we find $(p_k, \xi_k) \in \operatorname{gph} S$ such that

$$d((p_k, \xi_k), (\bar{p}, x_k))^q < \psi_S(x_k, \bar{p})^q + \delta_k < 2 \cdot \psi_S(x_k, \bar{p})^q =: b_k.$$

By definition of $d((p_k, \xi_k), (\bar{p}, x_k))$ thus also $d(\xi_k, x_k)^q < b_k$ and $\|p_k - \bar{p}\|^q < b_k$ hold true. In addition, the triangle inequality $\operatorname{dist}(x_k, S(\bar{p})) \leq d(x_k, \xi_k) + \operatorname{dist}(\xi_k, S(\bar{p}))$ yields

$$\operatorname{dist}(\xi_k, S(\bar{p})) \geq \operatorname{dist}(x_k, S(\bar{p})) - d(\xi_k, x_k) > \operatorname{dist}(x_k, S(\bar{p})) - b_k^{1/q}.$$

Note that, because of $b_k = o\big(\text{dist}(x_k, S(\bar{p}))\big)$ as well, we may choose in particular a subsequence s.t. $b_k < \text{dist}(x_k, S(\bar{p}))$ for all $k \in \mathbb{N}$.

As $0 < q \leq 1$ we get $b_k^{1/q} \leq b_k$ for $b_k \leq 1$ and so we obtain for $\xi_k \in S(p_k)$:

$$\frac{\|p_k - \bar{p}\|^q}{\text{dist}(\xi_k, S(\bar{p}))} < \frac{b_k}{\text{dist}(x_k, S(\bar{p})) - b_k^{1/q}} \leq \frac{b_k}{\text{dist}(x_k, S(\bar{p})) - b_k} \to 0,$$

as $k \to \infty$.

Hence, since $\xi_k \to \bar{x}$ and $\xi_k \in S(p_k)$ the map S cannot be calm $[q]$ at $(\bar{p}, \bar{x})$. □

Remark 15. Since it is $\text{dist}(x, S(\bar{p})) = 0 = \psi_S(x, \bar{p})$ for $x \in S(\bar{p})$, term (4.1) may be equivalently written as

$$\begin{aligned}\exists \varepsilon > 0\ \exists \alpha > 0\ \forall x \in \mathbb{B}(\bar{x}, \varepsilon):\\ x \notin S(\bar{p}) \to \alpha \, \text{dist}(x, S(\bar{p})) \leq \psi_S(x, \bar{p})^q,\end{aligned} \quad (4.3)$$

which is – by closedness of $\text{gph}\, S$ – the same as

$$\begin{aligned}\exists \varepsilon > 0\ \exists \alpha > 0\ \forall x \in \mathbb{B}(\bar{x}, \varepsilon):\\ \psi_S(x, \bar{p}) > 0 \to \alpha \, \text{dist}(x, S(\bar{p})) \leq \psi_S(x, \bar{p})^q.\end{aligned} \quad (4.4)$$

Finally, with any locally Lipschitzian function $\phi : X \to \mathbb{R}$ such that

$$\begin{aligned}c_1 \phi(x) \leq \psi_S(x, \bar{p}) \leq c_2 \phi(x) \text{ for } x \text{ near } \bar{x}\\ \text{and certain constants } 0 < c_1 \leq c_2,\end{aligned} \quad (4.5)$$

and with the sublevel set mapping

$$\Sigma(r) = \{\, x \in X \mid \phi(x) \leq r \,\}, \quad (4.6)$$

we can conclude that calmness $[q]$ for any closed multifunction coincides with calmness $[q]$ of a Lipschitzian inequality, i.e. it holds

4 Hölder calmness – conditions and characterizations

(cf. [Kum09, Corollary 2.3] as well)

Corollary 20. *A multifunction S is calm $[q]$ at $(\bar{p}, \bar{x}) \in gph\, S \subset P \times X$ if and only if the sublevel set map Σ of some Lipschitz function ϕ satisfying (4.5) is calm $[q]$ at $(0, \bar{x}) \in \mathbb{R} \times X$.*

Note. In particular, we may put $\phi(x) = \psi_S(x, \bar{p})$, so considering such calmness $[q]$ problems is an all-purpose tool.

Proof. (4.5) yields $\psi_S(x, \bar{p}) = 0 \iff \phi(x) = 0$, so $S(\bar{p}) = \Sigma(0)$ and $\mathrm{dist}(x, S(\bar{p})) = \mathrm{dist}(x, \Sigma(0))$. Thus one easily sees – using again (4.5) – that the equivalent condition (4.4) for calmness $[q]$ of S at $(\bar{p}, \bar{x})$ is equivalent to

$$\exists \varepsilon > 0, \alpha' > 0 \text{ s.t.}$$
$$\forall x \in \mathbb{B}(\bar{x}, \varepsilon) \text{ with } \phi(x) > 0 : \alpha' \mathrm{dist}(x, \Sigma(0)) \leq \phi(x)^q.$$

and we are done by Corollary 5. □

***Remark* 16.** Notice – using Corollary 5 and $S(0) = S_q(0)$ – that calmness $[q]$ of sublevel set maps $S(r) = \{\, x \in X \mid \phi(x) \leq r \,\}$ at $(0, \bar{x})$ coincides with proper calmness of same rank of the mapping

$$r \mapsto S_q(r) := \{\, x \in X \mid (\phi(x)^+)^q \leq r \,\}. \tag{4.7}$$

However in general $(\phi(\cdot)^+)^q$ is no longer locally Lipschitz even though ϕ was.

Corollary 21. *Let $g_i : X \to \mathbb{R}$, $i = 1, \ldots, m$, lower semicontinuous functions, $S(p) = \{\, x \in X \mid \wedge_{i=1}^m g_i(x) \leq p_i \,\}$ where $p = (p_1, \ldots, p_m) \in \mathbb{R}^m$, and $\bar{x} \in S(0)$.*

Then for $f := \max_{i=1,\ldots,m} g_i$ the sublevel set mapping $S_f(r) = \{\, x \mid f(x) \leq r \,\}$ is calm $[q]$ at $(0, \bar{x}) \in \mathbb{R} \times X$ with rank L on $\mathbb{B}(\bar{x}, \varepsilon)$ if and only if so is S at $(0, \bar{x}) \in \mathbb{R}^m \times X$.

Proof. First note that f as a max-function of l.s.c. functions is as well l.s.c. So by Lemma 4 calmness $[q]$ of S_f at $(0, \bar{x})$ means

$$\exists \varepsilon > 0 \ \exists L > 0 \ \forall x \in \mathbb{B}(\bar{x}, \varepsilon) : \operatorname{dist}(x, S_f(0)) \leq L(f(x)^+)^q.$$

Since $S_f(0) = S(0)$ we thus have

$$\operatorname{dist}(x, S(0)) \leq L(f(x)^+)^q = L(\max_{i=1,\ldots,m} g_i(x)^+)^q$$
$$\leq L(\max_{i=1,\ldots,m} |p_i|)^q = L\,\|p\|_\infty^q$$

whenever $x \in S(p) \cap \mathbb{B}(\bar{x}, \varepsilon)$.

On the other hand, if S is calm $[q]$ at $(0, \bar{x})$ we have by Lemma 19

$$\exists \varepsilon > 0 \ \exists L > 0 \ \forall x \in \mathbb{B}(\bar{x}, \varepsilon) : \operatorname{dist}(x, S(0)) \leq L\psi_S(x, 0)^q.$$

And since $\psi_S(x, 0) \leq \max_{i=1,\ldots,m} g_i(x)^+ = f(x)^+$ as well as $S_f(0) = S(0)$ this yields calmness $[q]$ of S_f. □

Remark 17. In particular the corollary yields:
$S(p) = \{\, x \in X \mid \bigwedge_{i=1}^m g_i(x) \leq p_i \,\}$ is calm $[q]$ at $(0, \bar{x}) \in \operatorname{gph} S$ if and only if $S_{f^+}(r) = \{\, x \mid f^+(x) \leq r \,\}$ is calm $[q]$ at $(0, \bar{x}) \in \operatorname{gph} S_{f^+}$ for $f^+(x) := \max_i g_i(x)^+$.

4.1 The basic theorem

Before we will present one theorem which plays a central role in our argumentation, we show two easy but nevertheless useful technical lemmas.

The first one is that calmness $[q]$ of sublevel set maps is stable under translation:

Lemma 22. *Let X be a Banach space and $g : X \to \mathbb{R}$ l.s.c. with $g(\bar{x}) = 0$. Further let $L > 0$, $q > 0$ and $\varepsilon \in (0, \infty]$ be given.*

4 Hölder calmness – conditions and characterizations

Then for the translation $f(x) := g(\bar{x} - x)$ and the sublevel sets $S_g(p) = \{\, x \in X \mid g(x) \leq p \,\}$ and $S_f(p) = \{\, x \in X \mid f(x) \leq p \,\}$ the following propositions are equivalent (here $\mathbb{B}(x, \infty) := X$):

(i) $\forall\, x \in \mathbb{B}(\bar{x}, \varepsilon) : \mathrm{dist}(x, S_g(0)) \leq L\, (g(x)^+)^q$;

(ii) $\forall\, x \in \mathbb{B}(0, \varepsilon) : \mathrm{dist}(x, S_f(0)) \leq L\, (f(x)^+)^q$.

By Corollary 5 this means in particular that $S_f(p)$ is calm $[q]$ at $(0, \bar{x})$ iff $S_g(p)$ is calm $[q]$ at $(0, 0)$ (without changing ε).

Proof. This holds true because $x \in \mathbb{B}(0, \varepsilon)$ iff $(\bar{x} - x) \in \mathbb{B}(\bar{x}, \varepsilon)$ and $\mathrm{dist}(\bar{x} - x, S_g(0)) = \mathrm{dist}(x, S_f(0))$ (and vice versa).[1] □

And secondly we give a condition under which one may extend calmness $[q]$ from subspaces to the whole space:

Lemma 23. *Let X be a Hilbert space and $g : X \to \mathbb{R}$ lower semicontinuous with $g(0) = 0$. Further let X_1 be a subspace of X such that for some constant $c > 0$ and for all $x_1 \in X_1$ and $x_2 \in X_2 := X_1^\perp$ holds*

1. $g(x_1) \leq 0 \implies g(x_1 + x_2) \leq 0$ and

2. $g(x_1) > 0 \implies c \cdot g(x_1 + x_2) \geq g(x_1)$.

Then (with distances induced by the scalar product) the property

$$\forall\, x_1 \in \mathbb{B}(0, \varepsilon) : \; \mathrm{dist}(x_1, S_{X_1}(0)) \leq L\, \big(g(x_1)^+\big)^q$$

yields

$$\forall\, x \in \mathbb{B}(0, \varepsilon) : \; \mathrm{dist}(x, S(0)) \leq L\, \big(g(x)^+\big)^q\, ;$$

where $L > 0$, $q > 0$, $\varepsilon \in (0, \infty]$ are constants, and $S_{X_1}(p) = \{\, x_1 \in X_1 \mid g(x_1) \leq p \,\}$ as well as $S(p) = \{\, x \in X \mid g(x) \leq p \,\}$ as usual.

---

[1]Note that $g(x) = f(\bar{x} - x)$ iff $f(x) = g(\bar{x} - x)$. And of course the translation f is also l.s.c since we have for l.s.c. functions g and continuous functions h that $g \circ h$ is l.s.c. as well.

4.1 The basic theorem

So in particular, under the given conditions, calmness [q] of S_{X_1} on X_1 at $(0,0)$ implies calmness [q] of S at $(0,0)$.

Note. The conditions 1. and 2. particularly mean that $g(x_1) \leq 0 \Leftrightarrow g(x_1 + x_2) \leq 0$ for all $x_1 \in X_1$ and $x_2 \in X_1^\perp$ and thus – since $g(0) = 0$ was assumed – moreover $g(x_2) \leq 0$ for every $x_2 \in X_1^\perp$, which is rather restrictive.

Nevertheless the lemma will prove usefull later for a calmness [1/2] result of quadratic polynomials (see Theorem 49).

Proof. Let X_1 as requested. Then by assumption it holds

$$\forall x_1 \in \mathbb{B}_{X_1}(0, \varepsilon) \setminus S_{X_1}(0)\ \exists x_1' \in S_{X_1}(0) : \|x_1 - x'\|_2 \leq Lg(x_1)^q.$$

Now let $z \in \mathbb{B}(0, \varepsilon) \subset X = X_1 \oplus X_1^\perp$. Then there are (unique) $x_1 \in X_1$ and $x_2 \in X_1^\perp$ such that $z = x_1 + x_2$ and $\|z\|_2^2 = \|x_1\|_2^2 + \|x_2\|_2^2$, which yields in particular $x_1 \in \mathbb{B}_{X_1}(0, \varepsilon)$.

Further if $g(z) > 0$ it is $g(x_1) > 0$, so there is some $x_1' \in S_{X_1}(0)$ as above. Now $x_1' \in S_{X_1}(0)$ means $g(x_1') \leq 0$ and thus by our premise $g(x_1' + x_2) \leq 0$, i.e. $z' = x_1' + x_2 \in S(0)$. So finally we get $\|z - z'\|_2 = \|x_1 - x_1'\|_2 \leq Lg(x_1)^q \leq Lc^q g(z)^q$. □

Both above lemmas will become useful later to prove some of the central conclusions in Chapter 5 (cf. Theorems 49 and 50).

The following theorem is Proposition 3.4 in [Kum09]. For the case of proper calmness ($q = 1$) such statements may be found in [KK09] as well. For sake of completeness (and as there is some slight change in the formulation) we will give a comprehensive proof of said proposition, adapting the ideas of [Kum09, Section 2.4].

Theorem 24 ([Kum09, Proposition 3.4]). *Let X be a Banach space and $g : X \to \mathbb{R}$ lower semicontinuous with $g(\bar{x}) = 0$.*

4 Hölder calmness – conditions and characterizations

Then the sublevel set map $S(p) = \{ x \in X \mid g(x) \leq p \}$ is calm $[q]$ at $(0, \bar{x})$ if and only if

$$\exists \varepsilon, \lambda > 0 \, \forall x \in \mathbb{B}(\bar{x}, \varepsilon) \backslash S(0) \, \exists x' \in X : \\ (g(x')^+)^q - g(x)^q < -\lambda \|x - x'\|; \tag{4.8}$$

or equivalently

$$\exists \varepsilon, \lambda > 0 \, \forall x \in \mathbb{B}(\bar{x}, \varepsilon) \backslash S(0) \, \exists x' \in X : \\ \|x - x'\| < \lambda^{-1} \left((g(x')^+)^q - g(x)^q \right). \tag{4.9}$$

Further, under condition (4.8), the rank of calmness $[q]$ obtained is λ^{-1} (or greater).

Proof.
($\Rightarrow$) S is calm $[q]$ with rank L means by Corollary 5 that there is some $\varepsilon > 0$ s.t.

$$\forall x \in \mathbb{B}(\bar{x}, \varepsilon) \backslash S(0) \, \exists x' \in S(0) : \, \|x - x'\| \leq L g(x)^q.$$

Fix any $x \in \mathbb{B}(\bar{x}, \varepsilon) \backslash S(0)$ and the associated x'. Then $g(x)^q > \lambda \|x - x'\|$ for every $0 < \lambda < L^{-1}$. As $g(x') \leq 0$, i.e. $g(x')^+ = 0$, this yields $(g(x')^+)^q - g(x)^q < -\lambda \|x - x'\|$.

($\Leftarrow$) Consider (4.8) and let some $\delta > 0$ with $\lambda^{-1} \delta^q \leq \frac{1}{2}\varepsilon$ and arbitrary $p \in \mathbb{B}(0, \delta)$ and $x \in S(p) \cap \mathbb{B}(\bar{x}, \frac{1}{2}\varepsilon)$ be given. For $p \leq 0$ trivially holds $\mathrm{dist}(x, S(0)) = 0 \leq L \cdot |p|^q$ for any $L > 0$, so regard $0 < p \leq \delta$.

Next we will, starting with $(p_1, x_1) := (p, x) \in \mathrm{gph}\, S$, inductively construct a sequence $\{(p_k, x_k)\}$ in $\mathrm{gph}\, S$ such that for all $k > 1$ hold the following properties

(i) $0 \leq p_k \leq p_{k-1}$

(ii) $p_k^q + \lambda \|x_k - x_{k-1}\| \leq p_{k-1}^q$

4.1 The basic theorem

(iii) $p_k^q + \lambda \|x_k - x_{k-1}\| < \mu(p_{k-1}, x_{k-1}) + \frac{1}{k-1}$

where

$$\mu(p, x) := \inf \{ \, \tilde{p}^q + \lambda \|\tilde{x} - x\| \mid (\tilde{p}, \tilde{x}) \in \text{gph}\, S \,\wedge\, 0 \leq \tilde{p} \leq p \, \}.$$

So let $(p_1, x_1), \ldots, (p_k, x_k) \in \text{gph}\, S$ with the above properties be given – at least the starting point (p_1, x_1) is given anyway – and construct (p_{k+1}, x_{k+1}):

First note that $x_1, \ldots, x_k \in \mathbb{B}(\bar{x}, \varepsilon)$ as $x_1 \in \mathbb{B}(\bar{x}, \frac{1}{2}\varepsilon)$ by assumption and

$$\|x_k - x_1\| \leq \sum_{j=1}^{k-1} \|x_{j+1} - x_j\| \underset{(ii)}{\leq} \lambda^{-1} \sum_{j=1}^{k-1} \left(p_j^q - p_{j+1}^q\right) \qquad (4.10)$$
$$= \lambda^{-1} \left(p_1^q - p_{k+1}^q\right) \leq \lambda^{-1} p_1^q \leq \lambda^{-1} \delta^q \leq \frac{1}{2}\varepsilon.$$

If now $0 < g(x_k)$ then by (4.8) there is some x' s.t.

$$(g(x')^+)^q + \lambda \|x_k - x'\| < g(x_k)^q. \qquad (4.11)$$

Thus in particular $0 \leq g(x')^+ < g(x_k) \leq p_k$. As $(g(x')^+, x') \in \text{gph}\, S$ thus $\mu(p_k, x_k)$ exists, so one finds a tuple $(p_{k+1}, x_{k+1}) \in \text{gph}\, S$ with $0 \leq p_{k+1} \leq p_k$ such that

$$p_{k+1}^q + \lambda \|x_{k+1} - x_k\| - \frac{1}{k} < \mu(p_k, x_k)$$

and

$$p_{k+1}^q + \lambda \|x_{k+1} - x_k\| \leq (g(x')^+)^q + \lambda \|x' - x_k\| < p_k^q.$$

For the case $g(x_k) \leq 0$ we just put $(p_{k+1}, x_{k+1}) = (0, x_k)$ which yields $p_{k+1}^q + \lambda \|x_{k+1} - x_k\| = 0$. In any case (p_{k+1}, x_{k+1}) fulfills the desired properties.

By (4.10) in particular we have $\sum_{j=1}^{k} \|x_{j+1} - x_j\| \leq \lambda^{-1} p_1^q$ for every $k \geq 1$, i.e.
$$\sum_{j=1}^{\infty} \|x_{j+1} - x_j\| \leq \lambda^{-1} p_1^q,$$
and thus $\{x_k\}$ is a Cauchy sequence which converges towards some ξ in the complete space X, moreover $\xi \in \mathbb{B}(\bar{x}, \varepsilon)$ as $\{x_k\} \subset \mathbb{B}(\bar{x}, \varepsilon)$. Additionally, as $0 \leq p_{k+1} \leq p_k$, there is some $\eta \geq 0$ with $p_k \searrow \eta$. And since S is closed, we have $(\eta, \xi) \in \operatorname{gph} S$ which means $g(\xi) \leq \eta$.

Now, for $g(\xi) \leq 0$ it holds
$$\operatorname{dist}(x, S(0)) \leq \|x - \xi\| = \|x_1 - \xi\| \leq \lambda^{-1} p_1^q = \lambda^{-1} p^q, \quad (4.12)$$
so we are done if $\eta = 0$.

Suppose $0 < g(\xi) \leq \eta$. Hence, using condition (4.8) as above, we find some $(\tilde{p}, \tilde{x}) \in \operatorname{gph} S$ with $0 \leq \tilde{p} \leq \eta$ such that $\tilde{p}^q + \lambda \|\tilde{x} - \xi\| < \eta^q$. In particular there exists $\alpha > 0$ s.t.
$$\tilde{p}^q + \lambda \|\tilde{x} - \xi\| < \eta^q - \alpha,$$
so $\|\tilde{x} - x_k\| \to \|\tilde{x} - \xi\|$ yields for k large enough that
$$\tilde{p}^q + \lambda \|\tilde{x} - x_k\| < \eta^q - \alpha. \quad (4.13)$$
On the other hand, since $0 \leq \tilde{p} \leq \eta \leq p_k$, it is
$$\mu(p_k, x_k) \leq \tilde{p}^q + \lambda \|\tilde{x} - x_k\| \quad (4.14)$$
for every k.

So (iii), (4.14) and (4.13) imply that
$$p_{k+1}^q \leq p_{k+1}^q + \lambda \|x_{k+1} - x_k\|$$
$$< \mu(p_k, x_k) + \frac{1}{k} < \eta^q - \alpha + \frac{1}{k} \leq \eta^q - \frac{\alpha}{2} < \eta^q$$

4.1 The basic theorem

for every large k. But this contradicts $0 \le \eta \le p_k$ for all k. □

Remark 18. Remark 6 indicates that it is sufficient to show that for every $x_k \to \bar{x}$ with $x_k \ne \bar{x}$, $0 < g(x_k)$ and $\lim_{k\to\infty} g(x_k)^q \|x_k - \bar{x}\|^{-1} = 0$ exists x'_k s.t. $(g(x'_k)^+)^q - g(x_k)^q < -\lambda \|x_k - x'\|$ in order to prove calmness [q] of sublevel set maps S as in Theorem 24.

Checking the proof of Theorem 24 one sees that there is a global variant of said theorem. Similar propositions were proved by Wu and Ye [WY02a, Theorem 5] and by Ng and Zheng [NZ00, Theorem 1, Corollary 1 and Corollary 2], where the latter needed some additional requirements (see as well [NZ01] for corresponding results for exponent $q = 1$).

Theorem 25. *Let again X be a Banach space, $g : X \to \mathbb{R}$ lower semicontinuous and $S(p) = \{\, x \in X \mid g(x) \le p \,\}$ where $S(0) \ne \emptyset$. Further let $L > 0$ and $q > 0$ be given such that*

$$\forall x \in X\backslash S(0) \; \exists x' \in X : \; \|x - x'\| < L\left(g(x)^q - (g(x')^+)^q\right). \quad (4.15)$$

Then
$$\forall x \in X : \; \mathrm{dist}(x, S(0)) \le L\left(g(x)^+\right)^q.$$

Note. There is also some kind of backward direction: Since $S(0)$ is closed, property (4.15) is true if $\mathrm{dist}(x, S(0)) < L\left(g(x)^+\right)^q$ for each $x \in X$.

Proof. Use the construction in the proof of Theorem 24 starting with $(p_1, x_1) := (g(x), x)$ if $x \in X \setminus S(0)$ and putting $\lambda = L^{-1}$. □

Remark 19. Theorem 25 is also true for level set maps $S(p) = \{\, x \mid h(x) = p \,\}$ where $h : \mathbb{R}^n \to \mathbb{R}$ is such that $|h|$ is a lower semicontinuous function[2]. Just put $g(x) := |h(x)|$ and note $S_g(p) := \{\, x \mid g(x) \le p \,\}$ is closed as well as $S_g(0) = S(0)$ to see this.

---
[2] For some example where $|h|$ is l.s.c. but h is not see for instance [WY02b, Example 3.1].

4.2 Iteration schemes for Hölder calmness

Klatte and Kummer [KK09] gave some algorithmic approach to describe Aubin property and calmness, which is partially extended in [Kum09] to the Hölder-type setting used here. We want to generalize the given procedures and algorithms to characterize calmness [q].

Remark 20. If one replaces "calmness [q]" with

$$\exists \varepsilon \in (0,\infty] \; \exists L > 0 \; \forall x \in \mathbb{B}(\bar{x},\varepsilon)\setminus S(0):$$
$$\mathrm{dist}(x, S(0)) \leq L \left(g(x)^+\right)^q,$$

then all propositions of this section remain true in their corresponding global form.

We start our examination of iteration schemes with the following remark:

Remark 21. The characteristic condition (4.8) for calmness [q] of a sublevel set is of course equivalent to

$$\exists \varepsilon, \lambda > 0 \; \forall x \in \mathbb{B}(\bar{x},\varepsilon)\setminus S(0) \; \exists u \in \mathbb{B}(0,1), \; t > 0:$$
$$\left(g(x+tu)^+\right)^q - g(x)^q < -\lambda t; \quad (4.16)$$

and the inequation $(g(x+tu)^+)^q - g(x)^q < -\lambda t$ implies in particular $t < \lambda^{-1} g(x)^q$.

Having this we can characterize calmness [q] using the following algorithm:

Algorithm 1. *Let $x_1 \in X$ and $\lambda_1 = 1$ be given.*
Step $k \geq 1$:

4.2 Iteration schemes for Hölder calmness

IF $g(x_k) \leq 0$ THEN put $(x_{k+1}, \lambda_{k+1}) = (x_k, \lambda_k)$;
ELSE find $u \in X$ and $t \in \mathbb{R}$ such that

$$\begin{aligned} &\|u\| \leq 1, \\ &0 < t \leq \lambda_k^{-1} g(x_k)^q, \\ &\left(g(x_k + tu)^+\right)^q - g(x_k)^q \leq -\lambda_k t. \end{aligned} \quad (4.17)$$

IF (u,t) exists THEN put $x_{k+1} = x_k + tu$ and $\lambda_{k+1} = \lambda_k$;
ELSE put $x_{k+1} = x_k$ and $\lambda_{k+1} = \frac{1}{2}\lambda_k$.
Put $k := k+1$ and repeat.

Lemma 26. *Let $g : X \to \mathbb{R}$ be continuous and $g(\bar{x}) = 0$. Then its sublevel set S is calm $[q]$ at $(0, \bar{x})$ if and only if exist $\alpha \in (0,1)$ and $\beta > 0$ such that $\lambda_k \geq \alpha$ holds for every step k of Algorithm 1 for each initial point satisfying $\|x_1 - \bar{x}\| \leq \beta$.*

In this situation, the sequence $\{x_k\}$ generated by Algorithm 1 converges to some $\xi \in S(0)$ and it holds

$$\|\xi - x_1\| \leq \alpha^{-1} \left(g(x_1)^+\right)^q. \quad (4.18)$$

Proof.
($\Leftarrow$) Having α, β as required, condition (4.16) is satisfied for $0 < \lambda < \alpha$ and every $x = x_1 \in \mathbb{B}(\bar{x}, \beta)$. Hence S has the claimed calmness property.

($\Rightarrow$) Consider $\lambda \in (0,1)$ and $\varepsilon > 0$ such that for all $x \in \mathbb{B}(\bar{x}, \varepsilon)$ with $g(x) > 0$ exist $u \in \mathbb{B}(0,1)$ and $t > 0$ satisfying (4.16). Then this condition holds in particular for the largest $\lambda_n := 2^{-n} \leq \lambda$. Additionally, by continuity of g and since $g(\bar{x}) = 0$, there exists some $0 < \beta < \frac{1}{2}\varepsilon$ such that $\lambda^{-1} g(x)^q \leq \frac{1}{2}\varepsilon$ for all $x \in \mathbb{B}(\bar{x}, \beta) \setminus S(0)$.

Thus, as shown in the sufficiency part of the proof to Theorem 24, the algorithm produces a sequence $\{(x_k, \lambda_k)\}$ with $\lambda_k \geq \lambda_n =: \alpha$ and $x_k \to \xi \in S(0)$ provided that $\|x_1 - \bar{x}\| \leq \beta$. The estimate (4.18) is then given by (4.12). $\square$

4 Hölder calmness – conditions and characterizations

A drawback of Algorithm 1 is that it might be difficult to find suitable u and t or to find out that they do not exist. In the following we will try to overcome this.

As above the proof of Theorem 27 below will show that, having a certain condition, one may, by the successive assignment $x \mapsto x' = x+tu$, construct a converging sequence which yields calmness [q] and vice versa.

Theorem 27. *Let* $g : X \to \mathbb{R}$ *be l.s.c. with* $g(\bar{x}) = 0$. *Then its sublevel set* S *is calm* [q] *at* $(0, \bar{x})$ *if and only if*

$$\begin{aligned}&\exists \lambda \in (0,1] \; \exists \varepsilon > 0 \; \forall x \in \mathbb{B}(\bar{x}, \varepsilon)\setminus S(0)\\&\quad \exists u \in \mathbb{B}(0,1),\, t > 0:\\&\quad g(x+tu) - g(x) \leq -\lambda g(x)\\&\quad \text{and} \quad t \leq \lambda^{-1} g(x)^q.\end{aligned} \qquad (4.19)$$

Note. It suffices to consider $\lambda \in (0,1)$ only in property (4.19), because if it holds with $\lambda \geq 1$ then $g(x+tu) - g(x) \leq -\lambda g(x)$ directly yields $g(x+tu) \leq 0$ and thus $\mathrm{dist}(x, S(0)) \leq t \leq g(x)^q$.

Proof.

($\Rightarrow$) Suppose calmness [q] of S with rank L. So (cf. Lemma 5), given $x \in \mathbb{B}(\bar{x}, \varepsilon) \setminus S(0)$ with $\varepsilon > 0$ sufficiently small, one finds $\xi \in S(0)$ such that $\|x - \xi\| \leq Lg(x)^q$.

Now put $t = \|\xi - x\|$ and $u = \frac{\xi - x}{t}$. Then $t \leq Lg(x)^q$ and hence also $t \leq \lambda^{-1} g(x)^q$ for any fixed $0 < \lambda \leq \min\{L^{-1}, 1\}$. And moreover, since $\lambda \leq 1$, we get

$$g(x+tu) - g(x) = g(\xi) - g(x) \leq -g(x) \leq -\lambda g(x).$$

($\Leftarrow$) Let $0 < \delta \leq \left(\frac{1}{2}\lambda\varepsilon(1 - (1-\lambda)^q)\right)^{1/q}$. Then for any $p \in \mathbb{B}(0,\delta)$ and each $x \in \mathbb{B}(\bar{x}, \frac{1}{2}\varepsilon) \cap S(p)$ with $g(x) > 0$ we argue as follows:

4.2 Iteration schemes for Hölder calmness

Set $\theta := 1 - \lambda \in (0, 1)$. Then, taking $x_1 := x$ as a starting point, we construct a sequence $\{x_k\} \subset \mathbb{B}(0, \varepsilon)$ with

$$g(x_{k+1}) \leq \theta g(x_k) \leq \theta^k g(x_1) \text{ and} \qquad (4.20)$$
$$0 < t_k \leq \lambda^{-1} g(x_k)^q \leq \lambda^{-1} \left(\theta^{k-1} g(x_1)\right)^q,$$

selecting related u_k and t_k from (4.19) and setting $x_{k+1} = x_k + t_k u_k$. Here (4.20) follows directly from (4.19) if $x_k \in \mathbb{B}(\bar{x}, \varepsilon)$ for each $k \in \mathbb{N}$. And this is ensured by

$$\|x_{k+1} - x_k\| \leq t_k \leq \lambda^{-1} \left(\theta^{k-1} g(x_1)\right)^q = \lambda^{-1} (\theta^q)^{k-1} g(x_1)^q$$

and

$$g(x_1) \leq p \leq \delta \leq \left(\frac{1}{2} \lambda \varepsilon (1 - \theta^q)\right)^{1/q},$$

which yield

$$\|x_{k+1} - x_1\| \leq \sum_{j=1}^{k} \|x_{j+1} - x_j\| \leq \lambda^{-1} g(x_1)^q \sum_{j=1}^{k} (\theta^q)^{j-1} \qquad (4.21)$$
$$\leq \lambda^{-1} g(x_1)^q \frac{1}{1 - \theta^q} \leq \frac{1}{2} \varepsilon,$$

so together with $x_1 \in \mathbb{B}(\bar{x}, \frac{1}{2}\varepsilon)$ it is $x_{k+1} \in \mathbb{B}(\bar{x}, \varepsilon)$ for each $k \in \mathbb{N}$.

Now (4.20) yields that t_k tends to zero and so the sequence $\{x_k\}$ is a Cauchy sequence in the complete space X, i.e. $\xi = \lim_{k \to \infty} x_k$ exists. So in particular $\liminf_{k \to \infty} g(x_k) = g(\xi)$, directly by lower semicontinuity of g. As by (4.20) it moreover is $\lim_{k \to \infty} g(x_k) = 0$, we get $\xi \in S(0)$.

So, checking (4.21), we finally have the Hölder estimate

$$\text{dist}(x, S(0)) \leq \|x - \xi\| = \|x_1 - \xi\| \leq \frac{1}{\lambda(1 - \theta^q)} g(x)^q, \qquad (4.22)$$

i.e. calmness $[q]$ holds. $\square$

4 Hölder calmness – conditions and characterizations

Note. For continuous g we do not need to consider $0 < \delta \le \left(\frac{1}{2}\lambda\varepsilon(1-(1-\lambda)^q)\right)^{1/q}$ (as in the above proof for the l.s.c. case), since – because of $g(\bar{x}) = 0$ – then exists some $0 < \varepsilon' < \frac{1}{2}\varepsilon$ such that

$$\lambda^{-1} g(x)^q \frac{1}{1-\theta^q} < \frac{1}{2}\varepsilon \quad \text{whenever} \quad \|x-\bar{x}\| \le \varepsilon'; \qquad (4.23)$$

so here we may restrict our reasoning to $x \in \mathbb{B}(\bar{x}, \varepsilon')$.

Remark 22 (Modification). If $(g^+)^q$ is Lipschitz continuous near $\bar{x}$ then one gets the additional estimate $\lambda g(x)^q \le t$ in (4.19), i.e. in this case the condition reads

$$\begin{aligned} \exists \lambda \in (0,1] \; \exists \varepsilon > 0 \; \forall x \in \mathbb{B}(\bar{x}, \varepsilon) \backslash S(0) \\ \exists u \in \mathbb{B}(0,1), \, t > 0 : \\ g(x+tu) - g(x) \le -\lambda g(x) \\ \text{and } \lambda g(x)^q \le t \le \lambda^{-1} g(x)^q, \end{aligned} \qquad (4.24)$$

by the following consideration:

Of course (4.24) is stronger than (4.19), so we only have to show that it is as well implied by calmness $[q]$ (under the given additional Lipschitz property): As in the necessity part of the proof of Theorem 27 we get for any given $x \in \mathbb{B}(\bar{x}, \varepsilon) \setminus S(0)$ some $u \in \mathbb{B}(0,1)$ and $t > 0$ s.t. $g(x+tu) = 0$, $g(x+tu) - g(x) \le -\lambda g(x)$ and $t \le \lambda^{-1} g(x)^q$.

Now let $C > 0$ be a Lipschitz constant of $(g^+)^q$ near $\bar{x}$. Then (decreasing ε if necessary to reach the zone of Lipschitz continuity), because $\lambda \in (0,1]$, it holds

$$\begin{aligned} 0 < \lambda g(x)^q &\le g(x)^q - g(x+tu)^q \\ &= |g(x)^q - g(x+tu)^q| \le Ct\|u\| \le Ct. \end{aligned}$$

This ensures $t \ge \frac{\lambda}{C} g(x)^q$, and so, after replacing λ with $\lambda' =$

4.2 Iteration schemes for Hölder calmness

$\min\{\lambda, C^{-1}\lambda\}$, property (4.19) plus $\lambda' g(x)^q \le t$ is satisfied. □

In particular, for (locally) Lipschitz g^+ condition (4.19) may be substituted for

$$\begin{aligned}
&\exists\, \lambda \in (0,1]\ \exists\, \varepsilon > 0\ \forall\, x \in \mathbb{B}(\bar{x}, \varepsilon) \setminus S(0) \\
&\quad \exists\, u \in \mathbb{B}(0,1),\ t > 0: \\
&\quad g(x + tu) - g(x) \le -\lambda g(x) \\
&\quad \text{and } \lambda g(x) \le t \le \lambda^{-1} g(x)^q.
\end{aligned} \qquad (4.25)$$

Remark 23 (Applying generalized derivatives). While the first condition of (4.19) is a usual descent condition, the second one does not appear in the context of known generalized derivatives or co-derivatives for (multi-)functions. It is needed to obtain a convergent sequence $\{x_k\}$.

So it may be not surprising that all sufficient (Hölder) calmness conditions (cf. also Section 4.5) based on known concepts of generalized (co-)derivatives for arbitrary functions or multifunctions, are not necessary – even for finite dimension.

In consequence of Theorem 27 calmness $[q]$ of sublevel sets (for l.s.c. functions) can be characterized by the following iteration scheme:

Algorithm 2. *Let $x_1 \in X$ and $\lambda_1 = 1$ be given.*
Step $k \ge 1$:
  IF $g(x_k) \le 0$ THEN put $(x_{k+1}, \lambda_{k+1}) = (x_k, \lambda_k)$;
  ELSE find $u \in X$ and $t \in \mathbb{R}$ such that

$$\begin{aligned}
&\|u\| \le 1, \\
&0 < t \le \lambda_k^{-1} g(x_k)^q, \\
&g(x_k + tu) - g(x_k) \le -\lambda_k g(x_k).
\end{aligned} \qquad (4.26)$$

IF (u,t) exists THEN put $x_{k+1} = x_k + tu$ and $\lambda_{k+1} = \lambda_k$;
ELSE put $x_{k+1} = x_k$ and $\lambda_{k+1} = \frac{1}{2}\lambda_k$.

Put $k := k+1$ *and repeat.*

Note. As noted above in Remark 22, one may replace (4.26) with

$$\begin{aligned}&\|u\| \leq 1,\\ &\lambda_k g(x_k) \leq t \leq \lambda_k^{-1} g(x_k)^q,\\ &g(x_k + tu) - g(x_k) \leq -\lambda_k g(x_k).\end{aligned} \quad (4.27)$$

in the case of (locally) Lipschitz functions.

Lemma 28. *Let* $g : X \to \mathbb{R}$ *be lower semicontinuous and* $g(\bar{x}) = 0$. *Then its sublevel set* S *is calm* $[q]$ *at* $(0, \bar{x})$ *if and only if there exist* $\alpha \in (0,1)$ *and* $\beta > 0$ *such that* $\lambda_k \geq \alpha$ *holds for every step* k *of Algorithm 2 whenever the initial point satisfies* $\|x_1 - \bar{x}\| \leq \beta$.

In this situation, the sequence $\{x_k\}$ *generated by Algorithm 2 converges to some* $\xi \in S(0)$ *and it holds*

$$g(x_{k+1}) \leq (1-\alpha)g(x_k) \quad \text{if } g(x_k) > 0 \quad \text{and}$$
$$\|\xi - x_1\| \leq \frac{(g(x_1)^+)^q}{\alpha(1-(1-\alpha)^q)}. \quad (4.28)$$

Proof. For numbers α and β fulfilling the desired properties, condition (4.19) is satisfied for each $0 < \lambda \leq \alpha$ and all $x = x_1 \in \mathbb{B}(\bar{x}, \beta)$, so we get calmness $[q]$ as claimed.

Conversely, assume that there are $\lambda \in (0,1)$ and $\varepsilon > 0$ such that, for all $x \in \mathbb{B}(\bar{x}, \varepsilon)$ with $g(x) > 0$ exist $u \in \mathbb{B}(0,1)$ and $t > 0$ satisfying (4.19). Then this also holds for the largest $\lambda_n = 2^{-n} \leq \lambda$. As shown in the sufficiency part of the proof to Theorem 27, thus the algorithm generates a sequence $\{(x_k, \lambda_k)\}$ with $\lambda_k \geq \lambda_n =: \alpha$ and $x_k \to \xi \in S(0)$ provided that $\|x_1 - \bar{x}\| \leq \varepsilon'$ is valid with $\beta := \varepsilon'$ from (4.23) (with λ_n instead of λ). The estimates (4.28) then follow from (4.20) and (4.22). □

Remark 24. If S is not calm $[q]$ at $(0, \bar{x})$, the sequence $\{\lambda_k\}$ obtained by Algorithm 2 obligatory tends to 0 (from above), so

4.2 Iteration schemes for Hölder calmness

one can choose a subsequence $\{\lambda_k'\}$ with $0 < \lambda_k' < \lambda_{k+1}'$ for each k. And because of $\bar{x} = x + \|\bar{x} - x\| \frac{\bar{x}-x}{\|\bar{x}-x\|}$ and ${\lambda_k'}^{-1} g(x)^q \to \infty$ for fixed x, it is not possible that $x_{k+1} = x_k$ for all k. Thus again and again there are x_k such that (4.26) has to be fulfilled.

As a result Algorithm 2 generates (in the non-calm $[q]$ case) a sequence containing a subsequence $\{(x_k, \lambda_k)\}$ satisfying $\lambda_{k+1} < \lambda_k$ and $x_{k+1} = x_k + tu$ with $g(x_{k+1}) < (1-\lambda_k)g(x_k)$ for some $u \in \mathbb{B}(0,1)$ and $t \in (0, \lambda_k^{-1} g(x_k)^q]$.

If we assume some more smoothness and regularity of the function in question, the next lemma yields another possible variation of Algorithm 2 and moreover a condition for Hölder calmness with exponent $\frac{1}{2}$.

Lemma 29. *Consider $g \in C^2(\mathbb{R}^n, \mathbb{R})$ with $g(\bar{x}) = 0$, $Dg(\bar{x}) = 0$ and regular Hessian $D^2 g(\bar{x})$. Then the steps of Algorithm 2 can be realized (for small λ_k) using any $u \in \mathbb{B}(0,1)$ with*

$$Dg(x_k) u \leq -\rho \|Dg(x_k)\|_2$$

for fixed $\rho \in (0,1)$ and setting $t = \lambda_k^{\frac{1}{2}} g(x_k)^{\frac{1}{2}}$.

In particular the given setting yields that the sublevel set S of g is calm $[1/2]$ at $(0, \bar{x})$.

Note. Corollary 46 below provides a different proof for calmness $[1/2]$ under the given assumptions.

Proof. To start with let us collect helpful properties following from the assumptions:

Since $D^2 g(\bar{x})$ is regular, we get, using the implicit function theorem, that the multifunction $\tilde{S} : \mathbb{R}^n \rightrightarrows \mathbb{R}^n$ defined as $\tilde{S}(y) := \{x \mid Dg(x) = y\}$ is pseudo-Lipschitz at $(0, \bar{x})$, i.e.

$$\exists C, \delta_1, \delta_2 > 0 \; \forall (\tilde{y}, \tilde{x}) \in (\mathbb{B}(0, \delta_1) \times \mathbb{B}(\bar{x}, \delta_2)) \cap \operatorname{gph} \tilde{S}$$
$$\forall y \in \mathbb{B}(0, \delta_1) \; \exists x \in \tilde{S}(y) : \|x - \tilde{x}\| \leq C \|y - \tilde{y}\|.$$

4 Hölder calmness – conditions and characterizations

And as Dg is continuous and $Dg(\bar{x}) = 0$ there exists some $\varepsilon > 0$ such that $Dg(x) \in \mathbb{B}(0, \delta_1)$ for all $x \in \mathbb{B}(\bar{x}, \varepsilon)$, so it follows

$$\forall\, x \in \mathbb{B}(\bar{x}, \varepsilon) : \|x - \bar{x}\| \leq C\|Dg(x)\|. \tag{4.29}$$

Further continuity of D^2g yields a constant $M > 0$ s.t. for all x in the compact set $\mathbb{B}(\bar{x}, \varepsilon)$

$$\|D^2g(x)\| \leq 2M,$$

and by Taylor's Theorem for each fixed $y \in \mathbb{R}^n$, $u \in \mathbb{B}(0, 1)$, $t \in \mathbb{R}$ it is

$$g(y + tu) = g(y) + tDg(y)u + \frac{t^2}{2}u^T D^2g(\xi)u$$

with $\xi = y + (1 - \vartheta)tu$ for some $\vartheta \in [0, 1]$.

Hence for all $x \in \mathbb{B}(\bar{x}, \varepsilon)$ we obtain: Whenever $x \notin S(0)$ then[3]

$$\begin{aligned} g(x) = |g(x) - g(\bar{x})| &= \left|\frac{1}{2}(x - \bar{x})^T D^2g(\xi)(x - \bar{x})\right| \\ &\leq \frac{1}{2}\|D^2g(\xi)\|\|x - \bar{x}\|^2 \leq M\|x - \bar{x}\|^2, \end{aligned} \tag{4.30}$$

and, for each fixed $u \in \mathbb{B}(0, 1)$, $t \in \mathbb{R}$ and $o_{u,x}(t) := \frac{t^2}{2}u^T D^2g(\xi)u$, it holds

$$\begin{aligned} g(x + tu) - g(x) &= tDg(x)u + o_{u,x}(t) \\ &\text{with}\quad |o_{u,x}(t)| \leq M\|u\|^2 t^2 \leq Mt^2. \end{aligned} \tag{4.31}$$

Next we will use the above to show that (4.26) is satisfiable using any $u \in \mathbb{B}(0, 1)$ with $Dg(x_k)u \leq -\rho\|Dg(x_k)\|$ and $t = \lambda_k^{\frac{1}{2}} g(x_k)^{\frac{1}{2}}$, where $\rho \in (0, 1)$ is fixed:

One directly gets $\lambda_k t = \lambda_k \lambda_k^{\frac{1}{2}} g(x_k)^{\frac{1}{2}} \leq g(x_k)^{\frac{1}{2}}$, because $\lambda_k \leq 1$. And using (4.31), we see that $g(x_k + tu) - g(x_k) \leq -\lambda_k g(x_k)$ is

---

[3] Remember that $g(\bar{x}) = 0$ and $Dg(\bar{x}) = 0$.

4.2 Iteration schemes for Hölder calmness

equivalent to

$$tDg(x_k)u + o_{u,x}(t) \leq -\lambda_k g(x_k),$$

which follows from

$$\lambda_k^{\frac{1}{2}} g(x_k)^{\frac{1}{2}} Dg(x_k)u + M\lambda_k g(x_k) \leq -\lambda_k g(x_k), \qquad (4.32)$$

if one chooses $t = \lambda_k^{\frac{1}{2}} g(x_k)^{\frac{1}{2}}$. Now (4.32) is equivalent to

$$\begin{aligned} Dg(x_k)u &\leq -\lambda_k^{\frac{1}{2}} g(x_k)^{\frac{1}{2}} - M\lambda_k^{\frac{1}{2}} g(x_k)^{\frac{1}{2}} \\ &= -(1+M)\lambda_k^{\frac{1}{2}} g(x_k)^{\frac{1}{2}}, \end{aligned} \qquad (4.33)$$

which, taking (4.30) into account, is true if the stronger condition

$$Dg(x_k)u \leq -M^{\frac{1}{2}}(1+M)\lambda_k^{\frac{1}{2}} \|x_k - \bar{x}\| \qquad (4.34)$$

holds. Since by (4.29) our specified u satisfies

$$Dg(x_k)u \leq -\rho \|Dg(x_k)\| \leq -\rho C^{-1} \|x_k - \bar{x}\|,$$

property (4.34) is fulfilled as soon as $\lambda_k \leq \rho^2 \left(C\sqrt{M}(1+M)\right)^{-2}$.

And because, for any fixed $\rho \in (0,1)$, with $u_k = -\frac{Dg(x_k)^T}{\|Dg(x_k)\|_2}$ it is

$$Dg(x_k)u_k \leq -\rho \|Dg(x_k)\|_2,$$

λ_k will not vanish and thus Lemma 28 yields calmness $[1/2]$ of S. Here we can consider $x_k \in \mathbb{B}(\bar{x}, \varepsilon)$ because of the following consideration:

Let $n := \min\left\{ k \in \mathbb{N} \mid 2^{-k} \leq \rho^2 \left(C\sqrt{M}(1+M)\right)^{-2} \right\}$. Then for $\lambda_n = 2^{-n}$ condition (4.34) holds true and so $\lambda_n = \lambda_k$ if $k \geq n$ and $\lambda_n < \lambda_k$ else.

Hence, with $\theta := (1 - \lambda_n)$, the sequence $\{x_k\}$ generated by the

4 Hölder calmness – conditions and characterizations

algorithm satisfies

$$g(x_{k+1}) \leq (1 - \lambda_k)g(x_k) \leq \theta g(x_k) \leq \theta^k g(x-1)$$

and therefore

$$\|x_{k+1} - x_1\| \leq \sum_{i=1}^{k-1} \|x_{i+1} - x_i\| \leq \sum_{i=1}^{k-1} \lambda_i^{\frac{1}{2}} g(x_i)^{\frac{1}{2}} \leq \sum_{i=1}^{k-1} g(x_i)^{\frac{1}{2}}$$

$$\leq g(x_1)^{\frac{1}{2}} \sum_{i=1}^{k-1} \left(\theta^{\frac{1}{2}}\right)^{i-1} \leq \frac{g(x_1)^{\frac{1}{2}}}{1 - \theta^{\frac{1}{2}}}.$$

This results in $x_k \in \mathbb{B}(\bar{x}, \varepsilon)$ whenever $x_1 \in \mathbb{B}(\bar{x}, \frac{1}{2}\varepsilon)$ and additionally x_1 is near enough to $\bar{x}$ s.t. $g(x_1) \leq (1 - \sqrt{1 - 2^{-n}})^2$ (which is possible because of continuity of g and $g(\bar{x}) = 0$). □

Remark 25 (Modification). In the lemma we needed $g \in C^2$ to obtain (4.29), (4.30) and (4.31). These properties are still ensured if Dg is locally Lipschitz, i.e. for so-called $C^{1,1}$ functions. Then (4.30) and (4.31) remain valid without additional assumptions and to get (4.29) it suffices to suppose that the contingent derivative $CDg(\bar{x})$ of Dg at $\bar{x}$ is injective, which replaces regularity of the Hessian $D^2 g(\bar{x})$.

Lemma 29 shows that the following algorithm may reveal proper calmness or calmness [1/2]:

Algorithm 3. *Let $x_1 \in \mathbb{R}^n$ and $\lambda_1 = 1$ be given.*
Step $k \geq 1$:
  IF $g(x_k) \leq 0$ THEN put $(x_{k+1}, \lambda_{k+1}) = (x_k, \lambda_k)$;
  ELSE
    IF $\|Dg(x_k)\|_2 \geq \gamma$ THEN find $u \in \mathbb{R}^n$ and $t \in \mathbb{R}$ such that

$$\begin{aligned} &\|u\| \leq 1, \\ &0 < t \leq \lambda_k^{-1} g(x_k), \\ &g(x_k + tu) - g(x_k) \leq -\lambda_k g(x_k). \end{aligned} \quad (4.35)$$

4.2 Iteration schemes for Hölder calmness

IF (u,t) *exists THEN put* $x_{k+1} = x_k + tu$ *and* $\lambda_{k+1} = \lambda_k$;
ELSE put $x_{k+1} = x_k$ *and* $\lambda_{k+1} = \frac{1}{2}\lambda_k$.
ELSE find $u \in \mathbb{R}^n$ *such that*

$$\begin{aligned}\|u\| &\leq 1,\\ Dg(x_k u)u &\leq -\rho \,\|Dg(x_k)\|_2\,.\end{aligned} \qquad (4.36)$$

IF u *exists*
THEN put $x_{k+1} = x_k + \sqrt{\lambda_k\, g(x_k)}\, u$ *and* $\lambda_{k+1} = \lambda_k$;
ELSE put $x_{k+1} = x_k$ *and* $\lambda_{k+1} = \frac{1}{2}\lambda_k$.
Put $k := k+1$ *and repeat.*

Note. Here γ and ρ are constants in $(0,1)$ to be chosen in advance.

As the proof of Theorem 33 will indicate, one could omit (4.36) and directly set $x_{k+1} = x_k - \sqrt{\lambda_k\, g(x_k)}\, \frac{Dg(x_k)^T}{\|Dg(x_k)\|_2}$ as soon as λ_k is small enough. This means we could replace the whole second part of Algorithm 3 with

ELSE
IF "λ_k *is small enough*"
THEN put $x_{k+1} = x_k - \sqrt{\lambda_k\, g(x_k)}\, \frac{Dg(x_k)^T}{\|Dg(x_k)\|_2}$ *and* $\lambda_{k+1} = \lambda_k$;
ELSE put $x_{k+1} = x_k$ *and* $\lambda_{k+1} = \frac{1}{2}\lambda_k$.

But how to check "λ_k is small enough"?

By condition (4.33) and taking $u = -\frac{Dg(x_k)^T}{\|Dg(x_k)\|_2}$ we could try to verify

$$\lambda_k \leq \frac{1}{(1+M)^2} \,\|Dg(x_k)\|_2^2\, g(x_k)^{-1},$$

but we do not know the constant M in advance ... so this seems to be a misleading idea.

On the basis of Theorem 27 we will next obtain the Hölder calm version of [KK09, Theorem 3] and then provide another algorithm to check calmness $[q]$.

4 Hölder calmness – conditions and characterizations

Theorem 30. *Let $q \in (0,1]$ and $g : X \to \mathbb{R}$ with $g(\bar{x}) = 0$ s.t. $(g^+)^q$ is Lipschitz near $\bar{x} \in X$. Then it holds:*

(i) *The sublevel set S of g is calm $[q]$ at $(0, \bar{x})$ if and only if there are $\lambda \in (0, 1]$, $\varepsilon > 0$ such that, for every $x \in \mathbb{B}(\bar{x}, \varepsilon)$ with $g(x) > 0$, there exist $u \in \mathbb{B}(0, 1)$ and $t > 0$ satisfying*

$$g(x + tu) - g(x) \leq -\lambda t^{\frac{1}{q}} \quad \text{and}$$
$$\lambda g(x)^q \leq t \leq \lambda^{-1} g(x)^q. \tag{4.37}$$

(ii) *Now let moreover g be given as $g(x) := \max_{i=1,\ldots,m} g_i(x)$ with $g_i \in C^1(X, \mathbb{R})$, $i = 1, \ldots, m$. As in [KK09], we define the relative slack of g_i as*

$$s_i(x) := \frac{g(x) - g_i(x)}{g(x)} \quad \text{if } g(x) > 0. \tag{4.38}$$

Then one may delete t and replace condition (4.37) with

$$Dg_i(x)u \leq \left(\frac{s_i(x)}{\lambda} - \lambda^{\frac{1}{q}}\right) g(x)^{1-q} \quad \forall\, i = 1, \ldots, m. \tag{4.39}$$

Note. For the case $q = 1$ the requested Lipschitz condition is of course automatically fulfilled if g is piecewise differentiable, but for $q \in (0, 1)$ we have to presume it.

Proof.

(i) We will show that condition (4.24) implies (4.37) and vice versa:

First note that $\lambda g(x)^q \leq t \leq \lambda^{-1} g(x)^q$ is equivalent to $\lambda^{\frac{1}{q}} g(x) \leq t^{\frac{1}{q}} \leq \lambda^{-\frac{1}{q}} g(x)$ and that – because $\lambda \in (0, 1]$ and $g(x) > 0$ – it is

$$\lambda^{1+\frac{1}{q}} g(x) \leq \lambda^{\frac{1}{q}} g(x) \quad \text{and} \quad \lambda^{-\frac{1}{q}} g(x) \leq \lambda^{-1-\frac{1}{q}} g(x).$$

4.2 Iteration schemes for Hölder calmness

Thus, whenever (4.24) or (4.37) hold, then also

$$\lambda^{1+\frac{1}{q}}g(x) \leq t^{\frac{1}{q}} \leq \lambda^{-1-\frac{1}{q}}g(x).$$

And since $\lambda^{\frac{1}{q}}g(x) \leq t^{\frac{1}{q}}$ yields $-\lambda t^{\frac{1}{q}} \leq -\lambda^{1+\frac{1}{q}}g(x)$ property (4.24) follows from (4.37) with new with $\lambda' = \lambda^{1+\frac{1}{q}}$. On the other hand $t^{\frac{1}{q}} \leq \lambda^{-\frac{1}{q}}g(x)$ gives $-\lambda g(x) \leq -\lambda^{1+\frac{1}{q}}t^{\frac{1}{q}}$, which shows that if (4.24) is true then (4.37) holds with $\lambda' = \lambda^{1+\frac{1}{q}}$.

Hence (4.24) and (4.37) are interchangeable, so we are done by Remark 22.

(ii) We start considering (4.37) and see that $g(x+tu) - g(x) \leq -\lambda t^{\frac{1}{q}}$ is for the given max-function g equivalent to

$$\forall i \in \{1,\ldots,m\}: \ g_i(x+tu) - g(x) \leq -\lambda t^{\frac{1}{q}},$$

which in turn is equivalent to

$$\forall i \in \{1,\ldots,m\}:$$
$$\frac{g_i(x+tu) - g_i(x)}{t} \leq \frac{g(x) - g_i(x)}{t} - \lambda t^{\frac{1}{q}-1}. \qquad (4.40)$$

Together with $\lambda g(x)^q \leq t$ this yields for each $i = 1,\ldots,m$

$$\begin{aligned}\frac{g_i(x+tu) - g_i(x)}{t} &\leq \frac{g(x) - g_i(x)}{\lambda g(x)^q} - \lambda(\lambda g(x)^q)^{\frac{1}{q}-1} \\ &= \frac{s_i(x)}{\lambda}g(x)^{1-q} - \lambda^{\frac{1}{q}}g(x)^{1-q}.\end{aligned} \qquad (4.41)$$

In addition it holds by assumption for all $i = 1,\ldots,m$

$$\lim_{t \to 0^+} \sup_{\substack{x \in B(\bar{x},\varepsilon) \\ u \in B(0,1)}} \left|\left(\frac{g_i(x+tu) - g_i(x)}{t} - Dg_i(x)u\right)(g(x)^+)^{q-1}\right| = 0,$$

i.e. the term $\frac{g_i(\cdot+tu)-g_i(\cdot)}{t}(g(\cdot)^+)^{q-1}$ converges uniformly to $Dg_i(\cdot)u\,(g(\cdot)^+)^{q-1}$. Thus, in particular, exist $\beta_i > 0$ s.t.

$$\forall\, t \in (0, \beta_i)\ \forall\, x \in \mathbb{B}(\bar{x}, \varepsilon)\ \forall\, u \in \mathbb{B}(0,1):$$
$$\left|\left(\frac{g_i(x+tu)-g_i(x)}{t} - Dg_i(x)u\right)(g(x)^+)^{q-1}\right| \leq \frac{1}{2}\lambda^{\frac{1}{q}}. \quad (4.42)$$

W.l.o.g. consider

$$\beta := \min_{i=1,\ldots,m} \beta_i \geq \lambda^{-1}\left(g(x)^+\right)^q$$

for all $x \in \mathbb{B}(\bar{x}, \varepsilon)$, which is possible because $g(\bar{x}) = 0$ and g is a continuous function.

Now take any $x \in \mathbb{B}(\bar{x}, \varepsilon)$ with $g(x) > 0$ and let u and t fulfill (4.37) for this x. So $t \leq \lambda^{-1}g(x)^q$, which yields with the above consideration $t \in (0, \beta)$ and hence

$$Dg_i(x)u \leq \frac{g_i(x+tu)-g_i(x)}{t} + \frac{1}{2}\lambda^{\frac{1}{q}}g(x)^{1-q}$$
$$\leq \left(\frac{s_i(x)}{\lambda} - \lambda^{\frac{1}{q}}\right)g(x)^{1-q} + \frac{1}{2}\lambda^{\frac{1}{q}}g(x)^{1-q}$$
$$= \left(\frac{s_i(x)}{\lambda} - \frac{1}{2}\lambda^{\frac{1}{q}}\right)g(x)^{1-q}$$
$$\leq \left(\frac{s_i(x)}{2^{-q}\lambda} - \left(\frac{\lambda}{2^q}\right)^{\frac{1}{q}}\right)g(x)^{1-q}.$$

So with $\lambda' = 2^{-q}\lambda$ instead of λ – and possibly new $\varepsilon > 0$ – we obtain (4.39).

Conversely, having (4.39), we conclude that (4.42) ensures

4.2 Iteration schemes for Hölder calmness

with $t = \frac{1}{2^q}\lambda g(x)^q$

$$\frac{g_i(x+tu) - g_i(x)}{t}$$
$$\leq Dg_i(x)u + \frac{1}{2}\lambda^{\frac{1}{q}}g(x)^{1-q}$$
$$\leq \left(\frac{s_i(x)}{2^{-q}\lambda} - \left(\frac{\lambda}{2^q}\right)^{\frac{1}{q}}\right)g(x)^{1-q}$$
$$= \left(\frac{g(x)^q}{t}s_i(x) - t^{\frac{1}{q}}g(x)^{-1}\right)g(x)^{1-q} \qquad (4.43)$$
$$= \frac{g(x) - g_i(x)}{t} - \frac{t^{\frac{1}{q}}}{g(x)^q}$$
$$= \frac{g(x) - g_i(x)}{t} - \frac{\lambda}{t}t^{\frac{1}{q}}$$
$$= \frac{g(x) - g_i(x)}{t} - \lambda t^{\frac{1}{q}-1}$$

for all $i \in \{1, \ldots, m\}$, i.e. (4.40) is fulfilled. $\square$

As a corollary of the Theorem we get in the case $q = 1$ the following alternative to (4.39) as stated in [KK09, Theorem 3]:

Corollary 31. Let $g := \max_{i=1,\ldots,m} g_i$ for $g_i \in C^1(X, \mathbb{R})$, $i = 1, \ldots, m$. Then the sublevel set S of g is calm at $(0, \bar{x}) \in gph\,S$ if and only if exist $\lambda \in (0, 1]$ and $\varepsilon > 0$ such that, for every $x \in \mathbb{B}(\bar{x}, \varepsilon)$ with $g(x) > 0$, there is some $u \in \mathbb{B}(0, 1)$ such that

$$Dg_i(\bar{x})u \leq \frac{s_i(x)}{\lambda} - \lambda =: b_i(x, \lambda) \qquad \forall i = 1, \ldots, m. \qquad (4.44)$$

Proof. We may choose $\varepsilon > 0$ such, that for all $x \in \mathbb{B}(\bar{x}, \varepsilon)$ and all $u \in \mathbb{B}(0, 1)$
$$|Dg_i(x)u - Dg_i(\bar{x})u| \leq \tfrac{1}{2}\lambda.$$

4 Hölder calmness – conditions and characterizations

So having calmness we get (4.39) with $q = 1$ and it follows

$$Dg_i(\bar{x})u \leq Dg_i(x)u + \tfrac{1}{2}\lambda \leq b_i(x,\lambda) + \tfrac{1}{2}\lambda \leq b_i(x,\tfrac{1}{2}\lambda).$$

The other direction is analogous. □

For the case $0 < q < 1$ the corresponding assertion to (4.44) is not true, i.e. it is not enough to have only

$$Dg_i(\bar{x})u \leq \left(\frac{s_i(x)}{\lambda} - \lambda^{\frac{1}{q}}\right) g(x)^{1-q} \quad \forall\, i = 1, \ldots, m,$$

for all x near $\bar{x}$ with fixed u and λ. We really need to consider $Dg_i(x)$:

Example 6. *Let $S(p) := \{\, x \in \mathbb{R} \mid g(x) := x^2 \leq p\,\}$. Then S is calm $[1/2]$ at $(0,0)$ (cf. Example 1), but*

$$g'(0)u = 0 > -\lambda^2 \sqrt{g(x)} \quad \forall\, \lambda > 0,\, x \neq 0,\, u \in [-1,1].$$

In contrast (4.37) and (4.39) hold with $\lambda = 1$, $u = \frac{x}{|x|}$ and $t = |x|$ for all $x \in \mathbb{R}\setminus\{0\}$.

The proof of Theorem 30 shows that it is sufficient to put $t = \frac{1}{2^q}\lambda\, g(x)^q$ – for $\lambda \in (0,1]$ small enough – in order to get property (4.39) whenever the sublevel set S of some piecewise differentiable function g is calm $[q]$ at $(0,\bar{x}) \in \operatorname{gph} S$.

As an important result we get, that Hölder calmness of such sets will be completely characterized by the following modified algorithm [KK09, ALG3], which uses the relative slack s_i (4.38) and the (computable) quantities

$$b_i^q(x,\lambda) := \left(\frac{s_i(x)}{\lambda} - \lambda^{\frac{1}{q}}\right) g(x)^{1-q} \tag{4.45}$$

for $g(x) > 0$ and $\lambda > 0$.

4.2 Iteration schemes for Hölder calmness

Note. Obviously, $b_i^q(x, \lambda)$ is increasing for $\lambda \to 0^+$ and fixed x.

Algorithm 4. *Let $x_1 \in X$ and $\lambda_1 = 1$ be given.*
Step $k \geq 1$:
  IF $g(x_k) \leq 0$ THEN put $(x_{k+1}, \lambda_{k+1}) = (x_k, \lambda_k)$;
  ELSE find $u \in X$ such that

$$\begin{aligned}&\|u\| \leq 1, \\ & Dg_i(x_k)u \leq b_i^q(x_k, \lambda_k) \quad \forall i \in \{1, \ldots, m\}.\end{aligned} \quad (4.46)$$

IF u exists
  THEN put $x_{k+1} = x_k + \frac{1}{2^q} \lambda_k \, g(x_k) u$ and $\lambda_{k+1} = \lambda_k$;
  ELSE put $x_{k+1} = x_k$ and $\lambda_{k+1} = \frac{1}{2}\lambda_k$.
Put $k := k + 1$ and repeat.

As before we have (see [KK09, Theorem 4 (ALG3)] for the genuine calm version of this result):

Lemma 32. *Let $g(x) := \max_{i=1,\ldots,m} g_i(x)$ with $g_i \in C^1(X, \mathbb{R})$, $i = 1, \ldots, m$, be such that $(g^+)^q$ is locally Lipschitz continuous near $\bar{x} \in X$.*

Then the sublevel set S of g is calm $[q]$ at $(0, \bar{x}) \in \operatorname{gph} S$ if and only if there exists some $\alpha > 0$ such that, for $\|x_1 - \bar{x}\|$ small enough, $\lambda_k \geq \alpha$ holds for every step k of Algorithm 4.

In this case, the sequence $\{x_k\}$ converges to some $\xi \in S(0)$ and it holds

$$g(x_{k+1}) \leq \left(1 - \tfrac{1}{2}\alpha^{1+\frac{1}{q}}\right) g(x_k) \quad \text{whenever } g(x_k) > 0. \quad (4.47)$$

Proof. The proof works as in Lemma 28, using Theorem 30 in addition to Theorem 27. And the estimate follows from (4.43), since this says $g(x_{k+1}) \leq g(x_k) - \alpha\, t^{\frac{1}{q}}$ and $t = \frac{1}{2^q} \alpha\, g(x_k)^q$. □

4 Hölder calmness – conditions and characterizations

Short note about solving the algorithms

Naturally the question arises how to find solutions to (4.17), (4.26) or (4.46). We will just have a short look at the convex system (4.46) in finite dimension, i.e. we want to find – for fixed x and λ – some $u \in \mathbb{R}^n$ such that

$$\begin{aligned} |u_j| \leq 1 \quad &\forall j \in J = \{1,\ldots,n\}, \\ Dg_i(x)u \leq b_i^q(x,\lambda) \quad &\forall i \in I = \{1,2,\ldots,m\}; \end{aligned} \qquad (4.48)$$

where $g_i \in C^1(\mathbb{R}^n, \mathbb{R})$. This is nothing more than to solve

$$\begin{aligned} |u_j| \leq 1 \quad &\forall j \in J = \{1,\ldots,n\}, \\ Au \leq b; \end{aligned} \qquad (4.49)$$

with $n \times m$ matrix A and $b \in \mathbb{R}^m$, which may be treated as any linear problem.

But let us check here whether it is suitable to use (a generalized version of) Newton's method. To do so consider functions $\psi : \mathbb{R}^m \to \mathbb{R}^m_+$ and $\varphi : \mathbb{R}^n \to \mathbb{R}^n_+$ defined by

$$\psi(v) = v^+ = (\max\{0, v_1\},\ldots,\max\{0, v_m\})^T$$

and

$$\varphi(u) = (\max\{0, |u_1|-1\},\ldots,\max\{0, |u_n|-1\})^T$$

respectively. Then (4.49) can be written as

$$G(u,v) := \begin{pmatrix} Au + \psi(v) - b \\ \varphi(u) \end{pmatrix} = 0. \qquad (4.50)$$

$G : \mathbb{R}^{n+m} \to \mathbb{R}^{n+m}$ is piecewise C^1, i.e. a PC^1-function, and thus semismooth. Additionally it is not only locally but globally Lipschitz.

Using that for PC^1-functions $f : \mathbb{R}^{d_1} \to \mathbb{R}^{d_2}$ with selection

4.2 Iteration schemes for Hölder calmness

functions f_k, $k = 1, \ldots, s$, by [Kum88a, Proposition 4] – see also [KK02b, p. 5], where as well the similar statement of Scholtes [Sch94, Proposition A.4.1] is listed – their Clarke differential $\partial^{Cl} f$ equals

$$\text{conv} \left\{ Df_k(x) \mid \exists x_l \to x : \left(f_k(x_l) = f(x_l) \wedge Df_k(x_l) = Df(x_l)\right) \right\},$$

we get

$$\partial^{Cl} G(u, v) = \left\{ \begin{pmatrix} A & B \\ C & 0 \end{pmatrix} \;\middle|\; B \in \partial^{Cl}\psi(v), C \in \partial^{Cl}\varphi(u) \right\}$$

with (e_i being the i-th unit vector)

$$\partial^{Cl}\psi_i(v) = \begin{cases} \{0\}, & v_i < 0 \\ \{e_i\}, & v_i > 0 \\ \{\lambda e_i \mid \lambda \in [0, 1]\}, & v_i = 0 \end{cases}$$

and

$$\partial^{Cl}\varphi_j(u) = \begin{cases} \{0\}, & -1 < u_j < 1 \\ \{e_j\}, & u_j > 1 \\ \{-e_j\}, & u_j < -1 \\ \{\lambda e_i \mid \lambda \in [0, 1]\}, & u_j = 1 \\ \{-\lambda e_i \mid \lambda \in [0, 1]\}, & u_j = -1. \end{cases}$$

As our point of interest is such that $-1 \leq u_j \leq 1$ for all $j = 1, \ldots, n$, it will be almost always the case that for some j it holds $\partial^{Cl}\varphi_j(u) = 0$ and thus the j-th line of $\partial^{Cl} G(u, v)$ will become zero. This means $\partial^{Cl} G$ is a singular matrix near most solutions and so Newton's method cannot work.

The only case such that $\partial^{Cl} G$ may be regular is that for the respective solution of (4.49) holds $|u_j| = 1$ for every $j = 1, \ldots, n$ and than the solution is easier to find without Newton's method.

4.3 Applying the algorithms

4.3.1 Arbitrary initial points

Until now we only analyzed the algorithms with respect to starting points near the points we are interested in. In the following let us check conditions and properties if we start at some arbitrary point.

Here we will focus on Algorithm 2 because it has a convenient testing condition which is nevertheless applicable for l.s.c. functions.

Theorem 33. *Let $g : X \to \mathbb{R}$ be l.s.c. on a reflexive Banach space X, $S(p) := \{\, x \mid g(x) \leq p \,\}$ its sublevel set and $x_1 \in X$ s.t. $S(g(x_1))$ is bounded.*

Then, for $0 < q \leq 1$, Algorithm 2 with starting point x_1 determines some $\xi \in S(0) \cap \mathbb{B}(x_1, L|g(x_1)|^q)$, if there is some some constant $\lambda > 0$ such that the sequence $\{\lambda_k\}$ generated by the algorithm satisfies $\lambda_k \geq \lambda$ for all $k \in \mathbb{N}$.

Otherwise, the occurring sequence $\{x_k\}$ has a weak accumulation point $\hat{x}$, which is stationary insofar that there are points z_k such that $\|z_k - x_k\| \to 0$ and

$$\liminf_{k \to \infty} \inf_{\|u\|=1} d^- g(z_k)(u) \geq 0. \tag{4.51}$$

Proof. The first part of the proposition holds, because the "⇐"-part of the proof of Theorem 27 shows, that (under the given assumption) $x_k \to \xi \in S(0)$ with

$$\|x_1 - \xi\| \leq \frac{1}{\lambda(1 - (1-\lambda)^q)} g(x_1)^q,$$

so in particular $L = (\lambda(1 - (1-\lambda)^q))^{-1}$.

4.3 Applying the algorithms

So assume $\lambda_k \to 0^+$. Thus, by condition (4.26), Algorithm 2 generates a sequence in X containing some subsequence $\{x_k\}$ with

$\forall\, u, t :$
$\left(\|u\| \leq 1 \wedge 0 < t \leq \lambda_k^{-1} g(x_k)^q \to (1 - \lambda_k)\, g(x_k) < g(x_k + tu) \right).$

So we have

$$\begin{aligned} 0 &< (1 - \lambda_k)\, g(x_k) \\ &\leq \inf \left\{ g(x) \mid \|x - x_k\| \leq \lambda_k^{-1} g(x_k)^q \right\} \\ &\leq g(x_k). \end{aligned} \quad (4.52)$$

Now we apply Ekeland's variational principle [Eke74] to the l.s.c. function g on the complete metric space $X_k := \mathbb{B}(x_k, \lambda_k^{-1} g(x_k)^q)$ on which g has a finite infimum:

By (4.52) it holds with $\varepsilon_E := \varepsilon_k = \lambda_k g(x_k)$ that $g(x_k) \leq \varepsilon_E + \inf_{x \in X_k} g(x)$, so for any $\alpha_E := \alpha_k = r_k \lambda_k^{-1} g(x_k)^q$, $r_k \in (0,1)$, there is some $z_k \in \mathbb{B}(x_k, \alpha_k) \subset \operatorname{int} X_k$ such that

$$(1 - \lambda_k)\, g(x_k) \leq g(z_k) \leq g(x_k) \quad (4.53)$$

and

$$\forall\, x' \in X_k : \; g(x') + \rho_k \|x' - z_k\| \geq g(z_k), \quad (4.54)$$

where $\rho_k := r_k^{-1} \lambda_k^2 g(x_k)^{1-q} = \frac{\varepsilon_E}{\alpha_E}$.

Setting $r_k = \lambda_k^{3/2}$ we obtain (recall $q \leq 1$)

$$\rho_k = \sqrt{\lambda_k}\, g(x_k)^{1-q} \leq \sqrt{\lambda_k}\, g(x_1)^{1-q} \xrightarrow[k \to \infty]{} 0^+,$$

and

$$\alpha_k = \sqrt{\lambda_k}\, g(x_k)^q \leq \sqrt{\lambda_k}\, g(x_1)^q \xrightarrow[k \to \infty]{} 0^+.$$

Since $\|x_k - z_k\| \leq \alpha_k$ this implies $\|x_k - z_k\| \to 0$. And as (4.54)

4 Hölder calmness – conditions and characterizations

yields for all $x' \in X_k$ that

$$-\rho_k \leq \frac{g(x') - g(z_k)}{\|x' - z_k\|} = \frac{g\left(z_k + \|x' - z_k\|\frac{x'-z_k}{\|x'-z_k\|}\right) - g(z_k)}{\|x' - z_k\|},$$

it follows

$$-\rho_k \leq \liminf_{\|x'-z_k\| \to 0^+} \frac{g\left(z_k + \|x' - z_k\|\frac{x'-z_k}{\|x'-z_k\|}\right) - g(z_k)}{\|x' - z_k\|}$$
$$= d^-g(z_k)\left(\frac{x' - z_k}{\|x' - z_k\|}\right)$$

for each $x' \in X_k$, and so

$$-\rho_k \leq \inf_{\|u\|=1} d^- g(z_k)(u), \qquad (4.55)$$

which results in (4.51) because $\rho_k \to 0$.

Since $S(g(x_1))$ is bounded and $g(z_k) \leq g(x_k) \leq g(x_1)$, i.e. $\{z_k\}, \{x_k\} \subset S(g(x_1))$, the sequences $\{z_k\}$ and $\{x_k\}$ are bounded. Thus, as X is reflexive, there exists a common[4] weak accumulation point $\hat{x} \in X$. □

Remark 26. If we know something more about the function g or its sublevel set S then we are able to say more about the weak accumulation point $\hat{x}$:

a) If g is weakly continuous, then $0 \leq g(\hat{x}) \leq g(x_1)$, so in particular $\hat{x}$ is contained in $S(g(x_1))$.

b) If the closed and bounded set $S(g(x_1))$ is weakly compact, then also $\hat{x} \in S(g(x_1))$.

---

[4]Remember $\|x_k - z_k\| \to 0$ and check the definition of weak convergence.

4.3 Applying the algorithms

c) For $X = \mathbb{R}^n$ the point $\hat{x}$ is a proper accumulation point of $\{x_k\}$.[5] Moreover, here the property $\|z_k - x_k\| \to 0$ yields $z_k \to \hat{x}$ for a subsequence.

d) If $g \in C^1(\mathbb{R}^n, \mathbb{R})$ then by (4.55) we have (with the same notation as in the above proof) for each $u \in \text{bd}\, \mathbb{B}(0,1)$ that

$$Dg(z_k)u \leq \rho_k,$$

and so with $u = \frac{Dg(z_k)^T}{\|Dg(z_k)\|_2}$ we get

$$\|Dg(z_k)\|_2 = Dg(z_k)u \leq \rho_k = \sqrt{\lambda_k} \xrightarrow[k \to \infty]{} 0. \qquad (4.56)$$

So because of continuity it is particularly $Dg(\hat{x}) = 0$ – which is what (4.51) means in the case of $g \in C^1(\mathbb{R}^n, \mathbb{R})$.

e) For the case that g is a C^1 maximum function, i.e. $g = \max_{i=1,\ldots,m} g_i$ with $g_i \in C^1(\mathbb{R}^n, \mathbb{R})$, proposition (4.51) implies

$$0 \in \partial^{Cl} g(\hat{x}),$$

which is equivalent to

$$\forall u \in \mathbb{R}^n : \max_{i \in I(\hat{x})} Dg_i(\hat{x})u \geq 0;$$

with $I(\hat{x})$ being the set of active indices at $\hat{x}$.

The boundedness assumption for $S(g(x_1))$ – which is nevertheless natural for many applications – cannot be omitted in general:

Example 7. *Consider $g(x) = e^x$, $q = 1$. Of course $S(g(x_1)) = (-\infty, x_1]$ is unbounded for any $x_1 \in \mathbb{R}$. And condition (4.26) of Algorithm 2 means to find in each step k some $\|u_k\| \leq 1$ and $t_k > 0$ such that $t_k \leq \lambda_k^{-1} e^{x_k}$ as well as $e^{t_k u_k} \leq 1 - \lambda_k$.*

---
[5] Recollect that in finite dimension weak convergence and convergence are the same.

4 Hölder calmness – conditions and characterizations

Now, if there was some $\lambda > 0$ with $\lambda_k \geq \lambda$ for all k, i.e. there are always $\|u_k\| \leq 1$, $t_k > 0$ satisfying $t_k \leq \lambda^{-1} e^{x_k}$ and $e^{t_k u_k} \leq 1 - \lambda$, then we would obtain[6]

$$t_k \leq \lambda^{-1} e^{x_1} \prod_{i=1}^{k-1} e^{t_i u_i} \leq \lambda^{-1} e^{x_1} (1-\lambda)^{k-1}.$$

But, as $0 < \lambda < 1$,[7] *this yields $|t_k u_k| \leq t_k \to 0^+$ and thus $e^{t_k u_k} \to 1$, which contradicts $e^{t_k u_k} \leq 1 - \lambda$.*

Hence Algorithm 2 generates a vanishing sequence λ_k and as in the proof of Theorem 33 exist z_k such that (4.54) is fulfilled. Particularly, if $r_k = \lambda_k^{3/2}$, this yields

$$e^{z_k} = g(z_k) = g'(z_k) = \lim_{h \to 0} \frac{e^{z_k+h} - e^{z^k}}{h}$$
$$= \lim_{h \to 0^-} \frac{e^{z^k} - e^{z_k+h}}{|h|} \leq \rho_k = \sqrt{\lambda_k},$$

and it follows $z_k \leq \ln(\sqrt{\lambda_k}) \to -\infty$. Since $|z_k - x_k| \to 0$, we thus also have $x_k \to -\infty$, so there is no (weak) accumulation point of $\{x_k\}$.

Remark 27. Clearly it holds that, if the assumptions of Theorem 33 are satisfied but a stationary weak accumulation point as mentioned cannot exist, then the sequence $\{\lambda_k\}$ cannot tend to zero and so Algorithm 2 determines necessarily some $\xi = \lim_{k \to \infty} x_k \in S(0) \cap \mathbb{B}(x_1, L|g(x_1)|^q)$.

This is for instance true if $g : \mathbb{R}^n \to \mathbb{R}$ is a convex function s.t. $\inf g < 0$ and $S(g(x_1))$ is bounded: Then (cf. Remark 26) we have $0 \leq g(\hat{x})$ and so for any $z_k \to \hat{x}$ it is $z_k \notin \arg\min g$ (at least for

---
[6]By construction it is $x_k = x_1 + \sum_{i=1}^{k-1} t_i u_i$.
[7]$\lambda = 1$ is not possible because there are no u and t s.t. $e^{x_1+tu} \leq (1-\lambda_1)e^{x_1} = 0$.

4.3 Applying the algorithms

large k). Here this means

$$0 \notin \partial g(z_k) = \{\, v \mid \forall\, w : \langle v, w \rangle \leq d^- g(z_k)(w)\,\},$$

i.e. there is some $w \in \mathbb{R}^n$ with $\|w\| = 1$ and $d^- g(z_k)(w) < 0$. So it follows

$$\liminf_{k \to \infty} \inf_{\|u\|=1} d^- g(z_k)(u) < 0.$$

4.3.2 Application to disturbed optimization problems

In this subsection we want to study the usage of the algorithms to find stationary points of classical nonlinear problems (NLP) in finite dimension[8]

$$\begin{aligned}&\min\{\,f(x) \mid g(x) \leq 0\,\}\\&\text{where } g = (g_1, \ldots, g_m) \text{ and } f, g_i \in C^1(\mathbb{R}^n, \mathbb{R}).\end{aligned} \quad (4.57)$$

We denote by $M := \{\, x \in \mathbb{R}^n \mid g(x) \leq 0\,\}$ the restriction set of (4.57) and call the function $L(x, y) := f(x) + \sum_{i=1}^m y_i g_i(x) = f(x) + \langle y, g(x) \rangle$ the *Lagrangian* of (4.57) where the components y_i of y are the *Lagrange multipliers*.

Definition 5. $(x, y) \in \mathbb{R}^{n+m}$ is called a *Karush-Kuhn-Tucker point* (short: KKT point) of (4.57) if it fulfills the *KKT conditions*:

1. $D_x L(x, y) = 0$ (Lagrange condition)

2. $g(x) \leq 0$ and $y \geq 0$ (Feasibility condition)

3. $\forall i\ y_i g_i(x) = 0$ (Complementarity condition)[9]

---

[8]We set equations aside as they here just make our considerations more laborious but not more meaningful.
[9]That is: $y_i \neq 0 \Rightarrow g_i(x) = 0$ and $g_i(x) \neq 0 \Rightarrow y_i = 0$.

4 Hölder calmness – conditions and characterizations

Let S_{KKT} be the *set of KKT points*, given by

$$\{\,(x,y) \mid D_x L(x,y) = 0 \wedge \forall\, i : \min\{y_i, -g_i(x)\} = 0\,\}. \quad (4.58)$$

If its set of Lagrange multipliers $Y(x) = \{\, y \mid (x,y) \in S_{\text{KKT}} \,\}$ is not empty then x is said to be a *stationary point*. Let S_{stat} denote the set of stationary points.

KKT points and their description

Let us first examine the relationship between Hölder calmness of a perturbed system describing KKT points and – on the other hand – of KKT points for perturbed optimization problems. To describe KKT points, we apply the simplest NCP function, the minimum of two arguments.

The perturbed system (4.58) (with $(a,b) \in \mathbb{R}^{n+m}$) is

$S_1(a,b) :=$
$\{\,(x,y) \mid D_x L(x,y) - a = 0,\ \min\{y_i, -g_i(x)\} - b_i = 0\ \forall i\,\}.$

By canonical perturbation of (4.57) we obtain

$$\min\{\, f(x) - \langle a, x \rangle \mid g(x) + b \leq 0 \,\} \quad (4.59)$$

(we changed the sign at b here) whose set of KKT points can be written as

$S_2(a,b) :=$
$\{\,(x,y) \mid D_x L(x,y) - a = 0,\ \min\{y_i, -(g_i(x) + b_i)\} = 0\ \forall i\,\}.$

Note. Obviously it holds $S_1(0,0) = S_{\text{KKT}} = S_2(0,0)$.

Below we will show that, though generally $S_1(a,b) \neq S_2(a,b)$ for $(a,b) \neq (0,0)$, calmness $[q]$ of S_1 and S_2 at the origin does

4.3 Applying the algorithms

not depend on the different descriptions. But let us start wit an example showing $S_1(a,b) \neq S_2(a,b)$ in general:

Example 8. *We consider $f(x) = x^2$ and $g(x) = x$. Then with $a = \frac{1}{2}$ and $b = -\frac{1}{2}$ we have $S_1(a,b) = \{(\frac{1}{2}, -\frac{1}{2})\}$ and $S_2(a,b) = \{(\frac{1}{4}, 0)\}$, because $D_x L(x,y) - a = 2x + y - \frac{1}{2}$ equals 0 iff $y = \frac{1}{2} - 2x$, and $\min\{\frac{1}{2} - 2x, -x\} = -\frac{1}{2}$ only holds for $x = \frac{1}{2}$ and $\min\{\frac{1}{2} - 2x, -x + \frac{1}{2}\} = 0$ is true for $x = \frac{1}{4}$ solely.*

Lemma 34. *Let $(\bar{x}, \bar{y})$ be a KKT point of (4.57). Then S_1 is calm [q] at $((0,0), (\bar{x}, \bar{y}))$ if and only if this is true for S_2.*

Proof. As we are in finite dimension, it suffices to prove the proposition for $\|\cdot\| = \|\cdot\|_\infty$.

($\Rightarrow$) Let the set S_1 be calm [q] at $((0,0), (\bar{x}, \bar{y}))$ with constants $\varepsilon_1, \delta_1, L_1 > 0$. We suppose w.l.o.g. $1 \geq \varepsilon_1 \geq 2\delta_1$.

As $g_i \in C^1$ it exists some $K \in \mathbb{R}$ such that for every x with $\|x - \bar{x}\| \leq \varepsilon_1$ it holds

$$\max_{1 \leq i \leq m} \|Dg_i(x)\| \leq K.$$

Furthermore if $(x, y) \in S_2(a, b)$, it is $(x, \tilde{y}) \in S_1(\tilde{a}, b)$ for

$$\tilde{y}_i = \begin{cases} b_i, & \text{if } y_i = 0 \text{ or } y_i < b_i \\ y_i, & \text{if } y_i > 0 \text{ and } y_i \geq b_i \end{cases} \quad \text{and} \quad \tilde{a} = D_x L(x, \tilde{y}),$$

and we have due to $y_i \geq 0$

$$\|y - \tilde{y}\| = \max_i |y_i - \tilde{y}_i| \leq \max_i |b_i| = \|b\|.$$

Then, by

$$\|a - \tilde{a}\| = \|D_x L(x, y) - D_x L(x, \tilde{y})\| = \left\|\sum_{1 \leq i \leq m}(y_i - \tilde{y}_i) Dg_i(x)\right\|$$
$$\leq m \|y - \tilde{y}\| \max_i \|Dg_i(x)\| \leq mK \|b\|,$$

also holds $\|\tilde{a}\| = \|\tilde{a} - a + a\| \leq \|\tilde{a} - a\| + \|a\| \leq mK\|b\| + \|a\| \leq (mK+1)\|(a,b)\|$.

Let now $\delta_2 := \frac{\delta_1}{mK+1}$, $\varepsilon_2 := \min\{\delta_2, \varepsilon_1\}$ and $L_2 := mKL_1 + L_1 + 1$. Then for every $(a,b) \in \mathbb{B}((0,0), \delta_2)$ and $(x,y) \in S_2(a,b) \cap \mathbb{B}((\bar{x}, \bar{y}), \varepsilon_2)$ it holds, with $\tilde{y}$ and $\tilde{a}$ as above,

$$\|(\tilde{a}, b)\| \leq (mK+1)\|(a,b)\| < (mK+1)\delta_2 = \delta_1,$$

and

$$\|(x, \tilde{y}) - (\bar{x}, \bar{y})\| \leq \|(x, \tilde{y}) - (x, y)\| + \|(x, y) - (\bar{x}, \bar{y})\|$$
$$\leq \|b\| + \varepsilon_2 \leq 2\delta_2 = \frac{2}{mK+1}\delta_1 < \varepsilon_1$$

and therefore (note that in particular $\|b\| \leq 1$ and $0 < q \leq 1$)

$$\operatorname{dist}((x,y), S_2(0,0)) \leq \|(x,y) - (x, \tilde{y})\| + \operatorname{dist}((x, \tilde{y}), S_2(0,0))$$
$$= \|y - \tilde{y}\| + \operatorname{dist}((x, \tilde{y}), S_1(0,0))$$
$$\leq \|b\| + L_1\|(\tilde{a}, b)\|^q$$
$$\leq \|b\|^q + L_1(mK+1)\|(a,b)\|^q$$
$$\leq L_2\|(a,b)\|^q.$$

($\Leftarrow$) We can do the same as above with

$$\tilde{y}_i = \begin{cases} 0, & \text{if } y_i = b_i \text{ or } y_i < 0 \\ y_i, & \text{if } y_i > b_i \text{ and } y_i \geq 0 \end{cases}.$$

$\square$

Computing stationary points

Here we assume $\min\{f(x) \mid g(x) \leq 0\} = 0$ for problem (4.57) and that $\bar{x}$ with $g(\bar{x}) = 0$ is a solution. Then of course also $\bar{x} \in S(0) := \{x \in \mathbb{R}^n \mid \wedge_{i=0}^m (g_i(x) \leq 0)\} = \{x \mid g^{\max}(x) \leq 0\}$, where $g_0 := f$ and $g^{\max} := \max_{i=0,\ldots,m} g_i$.

4.3 Applying the algorithms

Now we want to apply Algorithm 2 to

$$S(b) := \{\, x \mid g^{\max}(x) \leq 0 \,\},$$

presupposing that S is locally upper Lipschitz at $(0, \bar{x})$, i.e.

$$\begin{array}{l} \exists\, L,\, \beta,\, \delta > 0 \\ \forall\, b \in [0, \beta]\ \forall\, x \in S(b) \cap \mathbb{B}(\bar{x}, \delta) : \|x - \bar{x}\| \leq Lb. \end{array} \quad (4.60)$$

This assumption particularly yields that $\bar{x}$ is an isolated solution of $g^{\max}(x) \leq 0$, because $\forall\, x \in S(0) \cap \mathbb{B}(\bar{x}, \delta) : \|x - \bar{x}\| \leq L \cdot 0 = 0$.

For the application of the algorithm take a starting point x_1 near $\bar{x}$. We then have the following proposition:

Lemma 35. *Under the given assumptions Algorithm 2 generates a sequence $\{x_k\}$ such that $x_k \to \bar{x}$ or $x_k \to \hat{x}$ with*

$$0 \in \partial^{Cl} g^{\max}(\hat{x}), \quad (4.61)$$

if the starting point x_1 was sufficiently close to $\bar{x}$. Here $\hat{x}$ is a Fritz-John point of the parameterized problem (4.59) with $a = 0$ and $b = g^{max}(\hat{x})$.

If we further have the Mangasarian-Fromovitz constraint qualification (MFCQ) for the original problem at $\bar{x}$, and

$$\hat{x} = \hat{x}(x_1) \to \bar{x} \quad \text{for the starting point } x_1 \to \bar{x}, \quad (4.62)$$

then also $\hat{x} = \bar{x}$.

Proof. Directly by Theorem 33 (and the respective remarks from Remark 26) follows $x_k \to \xi \in S(0)$ or $x_k \to \hat{x}$ with $0 \in \partial^{Cl} g^{\max}(\hat{x})$. In the first case then isolatedness of $\bar{x}$ and the properties of the algorithm (cf. Lemma 28) ensure $\xi = \bar{x}$ for x_1 sufficiently close to $\bar{x}$.

Now by [Sch94, Proposition A.4.1] it is

$$\partial^{Cl} g^{\max}(\hat{x}) \subset \text{conv}\{ Dg_i(\hat{x}) \mid i \in I(\hat{x})\},$$

where $I(\hat{x}) = \{ i \in \{0, \ldots, m\} \mid g_i(\hat{x}) = g^{max}(\hat{x})\}$, and thus (4.61) yields the existence of some $\hat{y}_i \geq 0$, $i \in I(\hat{x})$, s.t. $\sum_{i \in I(\hat{x})} \hat{y}_i Dg_i(\hat{x}) = 0$ and $\sum_{i \in I(\hat{x})} \hat{y}_i = 1$. So, with $\hat{y}_i = 0$ if $i \notin I(\hat{x})$, we have $\sum_{i=0}^{m} \hat{y}_i Dg_i(\hat{x}) = 0$ as well as $\hat{y}_i \geq 0$, $g_i(\hat{x}) - b \leq 0$ and $(g_i(\hat{x}) - b)\hat{y}_i = 0$ for all $i = 1, \ldots, m$. But this means just by definition that $\hat{x}$ is a Fritz-John point of (4.59) with $a = 0$ and $b = g^{max}(\hat{x})$.

If moreover MFCQ holds, there is some $\bar{u} \in \mathbb{R}^n$ with $\|\bar{u}\| = 1$ such that

$$\forall\, i \in I(\bar{x}) : Dg_i(x)\bar{u} = \sum_{j=1}^{n} \partial_j g_i(x) \bar{u}_j < 0, \qquad (4.63)$$

whenever x is near enough to $\bar{x}$. So, if $\hat{x}$ is close to $\bar{x}$ (which we may assume by (4.62)) and thus fulfills (4.63), together with the Fritz-John property it follows

$$\begin{aligned}
0 &= \sum_{j=1}^{n} \sum_{i=0}^{m} \hat{y}_i \partial_j g_i(\hat{x}) \bar{u}_j \\
&= \sum_{i=1}^{m} \hat{y}_i \sum_{j=1}^{n} \partial_j g_i(\hat{x}) \bar{u}_j + \hat{y}_0 \sum_{j=1}^{n} \partial_j g_0(\hat{x}) \bar{u}_j \\
&< \hat{y}_0 \sum_{j=1}^{n} \partial_j g_0(\hat{x}) \bar{u}_j.
\end{aligned}$$

Hence in particular $\hat{y}_0 > 0$ and so $\hat{x}$ is stationary point for (4.59) with $a = 0$ and $b = g^{max}(\hat{x})$.

The upper Lipschitz property (4.60) now provides

$$\|\hat{x} - \bar{x}\| \leq Lb,$$

and thus we would have $\hat{x} = \bar{x}$, if $b = g^{max}(\hat{x}) = 0$.

4.3 Applying the algorithms

In order to show the latter, assume $b > 0$ and $\hat{x} \neq \bar{x}$. Since $\hat{y}_0 > 0$ additionally yields $g_0(\hat{x}) = g^{\max}(\hat{x})$, we obtain by the mean value theorem, that for any $\tilde{u} = \frac{\hat{x}-\bar{x}}{\|\hat{x}-\bar{x}\|}$ holds

$$b = g_0(\hat{x}) - g_0(\bar{x}) = Dg_0(\tilde{x})(\hat{x} - \bar{x}) = \tilde{t} Dg_0(\tilde{x})\tilde{u},$$

with $\tilde{t} = \|\hat{x} - \bar{x}\|$ and for some $\tilde{x}$ between $\hat{x}$ and $\bar{x}$. Because of $\tilde{t} \leq Lb$ it follows

$$Dg_0(\tilde{x})\tilde{u} = \frac{b}{\tilde{t}} \geq \frac{1}{L}.$$

As $\tilde{x} \to \bar{x}$ if $x_1 \to \bar{x}$ (remember that $\hat{x} = \hat{x}(x_1) \to \bar{x}$ and $\tilde{x}$ between $\hat{x}$ and $\bar{x}$), we thus get for every accumulation point $\hat{u}$ of the $\tilde{u}$ that

$$Dg_0(\bar{x})\hat{u} \geq \frac{1}{L}.$$

Further, since there is only a finite number of functions involved, we may pass to some subsequence of $x_1 \to \bar{x}$ providing us with constant sets $I(\hat{x}(x_1)) = I(\bar{x})$. Now, an analogous argument for the active constraints yields as well

$$Dg_i(\bar{x})\hat{u} \geq \frac{1}{L} \quad \text{if} \quad i \in I(\bar{x}).$$

But this means nothing else than $-\hat{u}$ being a descent direction at $\bar{x}$ for the active constraints and the objective function as well, and in consequence $\bar{x}$ cannot be a minimizer, a contradiction. $\square$

Remark 28. Lemma 35 states, that (under certain conditions) Algorithm 2 creates a sequence $\{x_k\}$ converging to $\bar{x}$ – whenever the starting point was close enough and provided that $\hat{x}$ remains close-by also.

But the latter is quite a problem: How can one guarantee that $\hat{x}$ stays close? Since, if we analyze Algorithm 2 (and the related propositions), we see that it is possible in the noncalm $[q]$ case, that we do not find u and t fulfilling (4.26) for a while and so x_k stays constant for the moment but λ_k^{-1} may become rather large – an then $t \le \lambda_k^{-1} g(x_k)^q$ may be such that $x_k + tu$ finally jumps away ...

Remark 29. If $\min\{f(x) \mid g(x) \le 0\} = v \ne 0$, then we may apply all the above to $g_0(x) := f(x) - v$.

Moreover, if the optimal value is not known but we have some lower bound $w \in \mathbb{R}$, then consider $g_0(x, x_{n+1}) := x_{n+1} - f(x)$, take

$$S(0) := \{\, x \in \mathbb{R}^{n+1} \mid g_i(x) \le 0;\ i = 0, \ldots, m \,\} = \{\, x \mid g^{\max}(x) \le 0 \,\}$$

and start the procedure near $(\bar{x}, f(\bar{x}))$.

4.4 Assigned linear inequality systems

In this chapter we again consider solution sets $S(p)$ for systems of inequalities of finitely many continuously differentiable functions g_i, $i = 1, \ldots, m$, or equivalently the sublevel set of the function f composed of the g_i, i.e. $f = \max_{i=1,\ldots,m} g_i$.

Since MFCQ yields calmness, a sufficient condition for calmness of S at some $(0, \bar{x}) \in \operatorname{gph} S$ is

$$\exists u \in \operatorname{bd} \mathbb{B}(0,1)\ \forall i \in I(\bar{x}) : Dg_i(\bar{x})u < 0,$$

where $I(\bar{x}) := \{\, i \mid g_i(\bar{x}) = 0 \,\}$ is the set of active indices in $\bar{x}$.

Remark 30. Henrion and Outrata also gave some sufficient condition of this type [HO05, Theorem 3]: *For $X = \mathbb{R}^n$, S is calm at $(0, \bar{x}) \in \operatorname{gph} S$ if, at $\bar{x}$, the Abadie CQ holds true and*

$$\exists u \in \operatorname{bd} \mathbb{B}(0,1)\ \forall i \in J : Dg_i(\bar{x})u < 0,$$

whenever J fulfills $J = \{\, i \mid g_i(\xi_k) = 0 \,\}$ for certain $\xi_k \to \bar{x}$, $\xi_k \in \operatorname{bd} S(0) \setminus \{\bar{x}\}$.

4.4 Assigned linear inequality systems

But, as noted in [Kum09, Remark 4.8], this sufficient condition is violated even for the linear system

$$S(p) = \{\, x \in \mathbb{R}^2 \mid g_1(x) := x_1 \leq p_1,\ g_2(x) := -x_1 \leq p_2 \,\}:$$

Since $S(0) = \{\,(0, x_2) \mid x_2 \in \mathbb{R}\,\} = \operatorname{bd} S(0)$ and $g_i(0, x_2) = 0$, $i = 1, 2$, it holds $J = \{1, 2\}$ for $\bar{x} = 0$. But $Dg_i(0) = \pm(1, 0)$, so $Dg_i(\bar{x})u$ cannot be less than zero for both indices and fixed u.

So what are the crucial index sets for calmness and moreover calmness $[q]$? To answer this question we introduce the following notation:

Notation. Let $F^+ := \{\, x \in X \mid f(x) > 0 \,\}$. We define, with $I = \{1, \ldots, m\}$,

$\Theta_0 :=$
$$\left\{ J \subset I \,\middle|\, \exists\,\{x_k\} \subset F^+ \colon \bigl(x_k \to \bar{x} \,\wedge\, \forall k \,\forall i \in J \colon g_i(x_k) = f(x_k)\bigr) \right\},$$

$\Theta_{\lim 0} :=$
$$\left\{ J \subset I \,\middle|\, \exists\,\{x_k\} \subset F^+ \colon \bigl(x_k \to \bar{x} \,\wedge\, \forall i \in J \colon \lim_{k \to \infty} s_i(x_k) = 0\bigr) \right\},$$

and

$$\Theta_\iota^{\max} := \left\{\, J \in \Theta_\iota \,\middle|\, \forall \tilde{J} \in \Theta_\iota \setminus \{J\} \colon J \not\subseteq \tilde{J} \,\right\}, \quad \iota \in \{0, \lim 0\}.$$

Note. All those sets are dependent on $\bar{x}$.

Θ_0 collects the index sets J of active functions g_i (i.e. $g_i = f$) for some sequence $x_k \xrightarrow{F^+} \bar{x}$. The index sets in $\Theta_{\lim 0}$ may additionally contain indices of almost active functions g_i, that means those with $\frac{g_i(x_k)}{f(x_k)} \to 1$.

And $\Theta_\iota^{\max}$ are the sets of the respective maximal index collections.

Remark 31. To begin with observe that – because there are only finitely many subsets J of $\{1, \ldots, m\}$ – for any sequence $x_k \to \bar{x}$

exists a subsequence $\{x_{k_l}\}$ such that for some J it is

$$I(x_{k_l}) := \{\, i \mid g_i(x_{k_l}) = f(x_{k_l})\,\} = J \quad \text{for all } l \in \mathbb{N}.$$

Thus in particular $\Theta_0 \neq \emptyset$ and hence also $\Theta_{\lim 0}$ as well as $\Theta_\iota^{\max}$ are nonempty. Furthermore obviously

$$\Theta_0 \subset \Theta_{\lim 0} \quad \text{and} \quad \Theta_\iota^{\max} \subset \Theta_\iota, \ \iota \in \{0, \lim 0\}.$$

And one easily verifies that $J \subset I(\bar{x})$ for all $J \in \Theta_{\lim 0}$[10], and further that $\tilde{J} \in \Theta_\iota$ whenever $\tilde{J} \subset J \in \Theta_\iota^{\max}$ as well as that for each $\tilde{J} \in \Theta_\iota$ there is one $J \in \Theta_\iota^{\max}$ containing $\tilde{J}$. Therefore it holds

$$\Theta_\iota = \{\, \tilde{J} \subset I(\bar{x}) \mid \exists J \in \Theta_\iota^{\max} : \ \tilde{J} \subset J \,\}.$$

Moreover, setting

$$I(x) := \{\, i \mid g_i(x) = f(x)\,\} \quad \text{and}$$
$$I_{\lim}(\{x_k\}) := \{\, i \mid \lim_{k \to \infty} s_i(x_k) = 0\,\} \quad \text{for } x_k \xrightarrow{F^+} \bar{x},$$

it is

$$\Theta_0^{\max} \subsetneq \Theta_0^I \subsetneq \Theta_0 \quad \text{and} \quad \Theta_{\lim 0}^{\max} \subsetneq \Theta_{\lim 0}^I \subsetneq \Theta_{\lim 0},$$

where

$$\Theta_0^I := \{\, J \subset I(\bar{x}) \mid \exists \{x_k\} \subset F^+ : \bigl(x_k \to \bar{x} \wedge \forall k : J = I(x_k)\bigr)\,\}$$

and

$$\Theta_{\lim 0}^I := \{\, J \subset I(\bar{x}) \mid \exists \{x_k\} \subset F^+ : \bigl(x_k \to \bar{x} \wedge J = I_{\lim}(\{x_k\})\bigr)\,\}.$$

One directly notes that $\emptyset \in \Theta_\iota$ but $\emptyset \notin \Theta_\iota^I$. Indeed we have even

---

[10] We have (for all $i \in J$) $0 = \lim_{k \to \infty} s_i(x_k) = \lim_{k \to \infty} \frac{f(x_k) - g_i(x_k)}{f(x_k)}$ which yields $f(x_k) - g_i(x_k) \to 0$ and so $g_i(x_k) \to f(\bar{x}) = 0$. Since also $g_i(x_k) \to g_i(\bar{x})$ (by continuity) we get $g_i(\bar{x}) = 0$ for all $i \in J$.

4.4 Assigned linear inequality systems

more, namely (even if one determines Θ_ι not to contain $\emptyset$) that all the sets defined above may be different:

Example 9.

a) Let $g_i : \mathbb{R} \to \mathbb{R}$, $i = 1, 2$, be given as

$$g_1(x) := \begin{cases} x^4 \sin(\frac{1}{x}) + x, & x \neq 0 \\ 0, & x = 0 \end{cases}$$

and

$$g_2(x) := \begin{cases} x^4 \sin(\frac{1}{2x}) + x, & x \neq 0 \\ 0, & x = 0 \end{cases},$$

which are C^1 functions.

With $f = \max\{g_1, g_2\}$ it now holds for $x_k = \frac{2}{k\pi}$ (with $n \in \mathbb{N}$ below)

$$f(x_k) = \begin{cases} g_1(x_k) = g_2(x_k), & k = 4n \\ g_1(x_k), & k = 8n+1 \vee 8n+5 \vee 8n+6 \\ g_2(x_k), & k = 8n+2 \vee 8n+3 \vee 8n+7 \end{cases}$$

and

$$I(x_k) = \begin{cases} \{1,2\}, & k = 4n \\ \{1\}, & k = 8n+1 \vee 8n+5 \vee 8n+6 \\ \{2\}, & k = 8n+2 \vee 8n+3 \vee 8n+7 \end{cases}.$$

Thus

$$\Theta_0 = \{\emptyset, \{1\}, \{2\}, \{1,2\}\},$$
$$\Theta_0^I = \{\{1\}, \{2\}, \{1,2\}\}, \text{ and}$$
$$\Theta_0^{\max} = \{\{1,2\}\}.$$

4 Hölder calmness – conditions and characterizations

b) *Next consider*

$$g_1(x) := \begin{cases} x^4 \sin(\frac{1}{x}) + x, & x \neq 0 \\ 0, & x = 0 \end{cases},$$

and

$$g_2(x) := x^4 + x.$$

Since $g_1(x) \leq g_2(x)$ for all $x \in \mathbb{R}$, the active index set $I(x)$ contains $i = 2$ for each sequence $x \to 0$ and therefore $\{1\} \notin \Theta_0^I$. But $\{1\} \in \Theta_0$ – using for instance $x_k = \frac{2}{(4k+1)\pi}$. Additionally here for all $x > 0$

$$|s_1(x)| = \frac{\left|\left(1 - \sin(\frac{1}{x})\right) x^4\right|}{x^4 + x} \leq \frac{2x^4}{x^4 + x} \xrightarrow[x \to 0]{} 0 \text{ and}$$

$$s_2(x) = 0,$$

so

$$\Theta_{\lim 0} = \{\emptyset, \{1\}, \{2\}, \{1,2\}\} \text{ and}$$
$$\Theta_{\lim 0}^I = \{\{1,2\}\} = \Theta_{\lim 0}^{\max}.$$

c) *Take $g_1(x) := x$ and $g_2(x) := \ln(x+1)$. As $\ln(x+1) < x$ for all $x > 0$ it is $I(x) = \{1\}$ for each $x > 0$ and hence $\Theta_0 = \{\emptyset, \{1\}\}$ and $\Theta_0^{\max} = \{\{1\}\} = \Theta_0^I$.*

On the other hand

$$\lim_{x \to 0^+} s_2(x) = \lim_{x \to 0^+} \frac{x - \ln(x+1)}{x} = \lim_{x \to 0^+} \frac{1 - \frac{1}{x+1}}{1} = 0,$$

which yields

$$\Theta_{\lim 0} = \{\emptyset, \{1\}, \{2\}, \{1,2\}\} \text{ and}$$
$$\Theta_{\lim 0}^{\max} = \{\{1,2\}\} = \Theta_{\lim 0}^I.$$

4.4 Assigned linear inequality systems

d) Finally with

$$g_1(x) := x^4 \text{ and}$$
$$g_2(x) := \frac{\left(1 - \cos(\frac{1}{x})\right) x^4}{4} + \frac{\left(1 + \cos(\frac{1}{x})\right) x^3 \ln(x+1)}{2},$$

one has for $x_k = \frac{1}{2k\pi}$ *that* $g_2(x_k) = x_k^3 \ln(x_k + 1) < x_k^4 = g_1(x_k)$ *and thus*

$$s_1(x_k) = 0 \quad \text{and} \quad s_2(x_k) = \frac{x_k - \ln(x_k + 1)}{x_k} \xrightarrow[x \to 0]{} 0,$$

i.e. $\Theta_{\lim 0}^{\max} = \{\{1, 2\}\}$. *But for* $x_k = \frac{1}{(2k+1)\pi}$ *it is* $g_2(x_k) = \frac{1}{2}x_k^4 < x_k^4 = g_1(x_k)$, *so*

$$s_1(x_k) = 0 \quad \text{and} \quad s_2(x_k) = \frac{1}{2},$$

and therefore $\Theta_{\lim 0}^I \supseteq \{\{1\}, \{1, 2\}\}$.

Notation. For $q \in (0, 1]$ we call a sequence $\{(x_k, \lambda_k)\} \subset F^+ \times \mathbb{R}^+$ *q-critical* with respect to $J \subset \{1, \ldots, m\}$ iff $x_k \to \bar{x}$, $\lambda_k \to 0^+$ (for $k \to \infty$) and

$$\forall i \in J : b_i^q(x_k, \lambda_k) := \left(\frac{s_i(x_k)}{\lambda_k} - \lambda_k^{\frac{1}{q}}\right) g(x_k)^{1-q} \xrightarrow[k \to \infty]{} 0^-.$$

In the particular case $q = 1$ we also use the name *critical* (instead of 1-critical) and denote $b_i(x_k, \lambda_k) := b_i^1(x_k, \lambda_k) = \lambda_k^{-1} s_i(x_k) - \lambda_k$.

Lemma 36. *For each* $q \in (0, 1]$ *and any* $J \in \Theta_{\lim 0}$ *exists a q-critical sequence* $\{(x_k, \lambda_k)\}$.

Proof. Take any $\{x_k\} \subset F^+$ with $x_k \to \bar{x}$ as well as $\forall i \in J : \lim_{k \to \infty} s_i(x_k) = 0$ and define $\mu_J(x) := \max_{i \in J} s_i(x)$ for $x \in F^+$.

4 Hölder calmness – conditions and characterizations

If $\mu_J(x_k) = 0$ for all $k \in \mathbb{N}$ then obviously for any sequence $\lambda_k \to 0^+$ it is $\{(x_k, \lambda_k)\}$ q-critical. So consider otherwise and put $\lambda_k := 2\,\mu_J(x_k)^{q^2/2}$. Then for all $i \in J$

$$b_i^q(x_k, \lambda_k) = \left(\lambda_k^{-1} s_i(x_k) - \lambda_k^{\frac{1}{q}}\right) g(x_k)^{1-q}$$
$$\leq \left(\frac{1}{2} s_i(x_k)^{1-\frac{q^2}{2}} - \lambda_k^{\frac{1}{q}}\right) g(x_k)^{1-q} \xrightarrow[k\to\infty]{} 0,$$

since $s_i(x_k) \to 0$, and thus $\lambda_k \to 0$, as well es $g(x_k)^{1-q} \to 0$ for $q \in (0,1)$ and $g(x_k)^{1-q} = 1$ for $q = 1$.

Moreover for k large enough it is $0 < \mu_J(x_k) \leq 1$ and thus $\mu_J(x_k)^{\frac{q}{2}} \geq \mu_J(x_k)^{\frac{1}{2}} \geq \mu_J(x_k)^{1-\frac{q^2}{2}}$, which yields

$$b_i^q(x_k, \lambda_k) \leq \left(\frac{1}{2} s_i(x_k)^{1-\frac{q^2}{2}} - 2^{\frac{1}{q}} \mu_J(x_k)^{\frac{q}{2}}\right) g(x_k)^{1-q}$$
$$\leq \left(\frac{1}{2} \mu_J(x_k)^{1-\frac{q^2}{2}} - 2^{\frac{1}{q}} \mu_J(x_k)^{\frac{q}{2}}\right) g(x_k)^{1-q}$$
$$\leq \left(\frac{1}{2}\sqrt{\mu_J(x_k)} - \sqrt{\mu_J(x_k)}\right) g(x_k)^{1-q}$$
$$= -\frac{1}{2}\sqrt{\mu_J(x_k)} g(x_k)^{1-q} < 0.$$

Thus for any $J \in \Theta_{\lim 0}$ one finds a q-critical sequence $\{(x_k, \lambda_k)\}$. □

Remark 32. Lemma 36 should have served as an entry point to generalize the statement of Theorem 37 below for Hölder calmness (which now is only a compilation of older results concerning proper calmness). But this approach failed as is noted in Remark 33. Nevertheless it yields some hints about which indices may be the the important ones regarding Algorithm 4.

Next we want to show that the above sets of critical indices (which are in general not only formally different as demonstrated

4.4 Assigned linear inequality systems

in Example 9) are playing the essential role for proper calmness of the sublevel set of a max-function composed of finitely many continuously differentiable functions. Here, the statement $(i) \Leftrightarrow (iii)$ is [Kum09, Theorem 4.7] and $(i) \Leftrightarrow (ii)$ may be found also in [HK06, Theorem 4.6].

Theorem 37. Let $g_i \in C^1(X, \mathbb{R})$,

$$S(p) = \{\, x \in X \mid \max_{i \in I} g_i(x) \leq p \,\}$$

and $(0, \bar{x}) \in gph\, S$. Then the following propositions are equivalent:

(i) S is calm at $(0, \bar{x})$;

(ii) $\forall J \in \Theta_{\lim 0}^{\max} \; \exists u \in \mathbb{B}(0,1) \; \forall i \in J: \; Dg_i(\bar{x})u < 0$;

(iii) $\forall J \in \Theta_0^{\max} \; \exists u \in \mathbb{B}(0,1) \; \forall i \in J: \; Dg_i(\bar{x})u < 0$.

Note. Of course we may assume $\bar{x} \in \text{bd}\, S(0)$, because else there is no sequence $x_k \xrightarrow{F^+} \bar{x}$ and thus $\Theta_{\lim 0} = \emptyset$ and $\Theta_0 = \emptyset$, so also the sets $\Theta_i^{\max}$ are empty. Thus (ii) and (iii) are trivially fulfilled – as well as calmness of S at inner points.

Proof.
$(i) \Rightarrow (ii)$ By Corollary 31 calmness yields

$$\exists \lambda, \varepsilon > 0 \; \forall x \in \mathbb{B}(\bar{x}, \varepsilon) \setminus S(0) \; \exists u \in \mathbb{B}(0,1)$$
$$\forall i = 1, \ldots, m: \; Dg_i(\bar{x})u \leq b_i(x, \lambda).$$

Now take any $J \in \Theta_{\lim 0}^{\max}$. By Lemma 36 we have a critical sequence $\{(x_k, \lambda_k)\}$ for J and w.l.o.g. we may suppose $\{x_k\} \subset \mathbb{B}(\bar{x}, \varepsilon) \setminus S(0)$ and $\lambda_k \leq \lambda$ for all k. Thus we have

$$\forall k \in \mathbb{N} \; \exists u \in \mathbb{B}(0,1) \; \forall i \in J:$$
$$Dg_i(\bar{x})u \leq b_i(x_k, \lambda) \leq b_i(x_k, \lambda_k) < 0.$$

4 Hölder calmness – conditions and characterizations

$(ii) \Rightarrow (iii)$ If for $J \in \Theta_{\lim 0}^{\max}$ exists $u \in \mathbb{B}(0,1)$ s.t. $\forall i \in J : Dg_i(\bar{x})u < 0$, then (with the same u) this holds for each subset of J – which are elements of $\Theta_{\lim 0}$. Thus, because of $\Theta_\iota = \{ \tilde{J} \subset I(\bar{x}) \mid \exists J \in \Theta_\iota^{\max} : \tilde{J} \subset J \}$, we have

$$\forall J \in \Theta_{\lim 0} \, \exists u \in \mathbb{B}(0,1) \, \forall i \in J : Dg_i(\bar{x})u < 0.$$

And since $\Theta_0^{\max} \subset \Theta_{\lim 0}$ we are done.

$(iii) \Rightarrow (i)$ Put $\lambda := -\frac{1}{2} \max_{J \in \Theta_0^{\max}} \max_{i \in J} Dg_i(\bar{x})u_J$, which exists since the set of indices is finite and which is by assumption greater than zero (we took u_J's with $\forall i \in J: Dg_i(\bar{x})u_J < 0$). Now let $\varepsilon > 0$ s.t. for all $x \in \mathbb{B}(\bar{x}, \varepsilon)$ holds $I(x) \in \Theta_0$ (this is possible because there are only finitely many $J \in \Theta_0$)[11] and

$$\forall J \in \Theta_0 \, \forall i \in J : |Dg_i(x)u_J - Dg_i(\bar{x})u_J| < \frac{1}{2}\lambda.$$

So, for each $x \in \mathbb{B}(\bar{x}, \varepsilon)$ and $t > 0$ small enough, it holds for at least one $i \in I(x)$

$$\frac{f(x + tu_{J_x}) - f(x)}{t} = \frac{g_i(x + tu_{J_x}) - g_i(x)}{t}$$
$$= Dg_i(x)u_{J_x} + o(t)$$
$$< Dg_i(\bar{x})u_{J_x} + \frac{1}{2}\lambda + o(t)$$
$$\leq -2\lambda + \frac{1}{2}\lambda + o(t)$$
$$\leq -\lambda,$$

i.e. $\frac{f(x+tu_{J_x})-f(x)}{t} < -\lambda$ for small $t > 0$ which yields calmness by Theorem 24. □

---

[11]Suppose not, then exists $x_k \xrightarrow{F^+} \bar{x}$ s.t. $\forall k \in \mathbb{N} : I(x_k) \notin \Theta_0$. Since there are only finitely many possible elements in Θ_0 there must be some subsequence $\{x_{k_l}\}$ s.t. $I(x_{k_l})$ is equal for all l. But this means $I(x_{k_l}) \in \Theta_0$, a contradiction.

Remark 33. In the same way as in the »$(i) \Rightarrow (ii)$«-part of above proof one gets – using Theorem 30 – that

$$\forall J \in \Theta_{\lim 0}^{\max} \exists x_k \to \bar{x}, s_i(x_k) \to 0 \ \exists u_k \in \mathbb{B}(0,1)$$
$$\forall i \in J : \ Dg_i(x_k)u_k < 0$$

is necessary for calmness $[q]$ at $(0,\bar{x}) \in \operatorname{gph} S$, which means nothing more than that in any neighbourhood of $\bar{x}$ there is a descent direction.

Of course this condition can not be sufficient: Since it does not take q into account, one could deduce that Hölder calmness for any $q_1, q_2 \in (0,1)$ is equivalent – and this is obviously not true.

4.5 Sufficient conditions

The next statement is well known (since the given requirements mean that LICQ is fulfilled, which is a stronger constraint qualification than calmness) but we want to give a proof using Theorem 24 as this will carry us to some generalization.

Corollary 38. Let X be a (real) subspace of $\mathbb{R}^n$ and $g \in C^1(X, \mathbb{R})$ such that $g(\bar{x}) = 0$ and $Dg(\bar{x}) \neq 0$ for some $\bar{x} \in X$. Then $S(p) := \{x \in X \mid g(x) \leq p\}$ is calm on X at $(0, \bar{x})$.

Proof. By continuity of Dg and as $Dg(\bar{x}) \neq 0$ there is some $\varepsilon > 0$ s.t. $Dg(x) \neq 0$ for all $x \in \mathbb{B}_X(\bar{x}, \varepsilon)$ and thus

$$\lambda := \min_{x \in \mathbb{B}_X(\bar{x},\varepsilon)} \max_{i=1,\ldots,n} |\partial_i g(x)| > 0.$$

This yields

$$\max_{\|u\|=1} |Dg(x)u| \geq \max_{i=1,\ldots,n} |\partial_i g(x)| \geq \lambda,$$

4 Hölder calmness – conditions and characterizations

so there is some $u_x \in \operatorname{bd} \mathbb{B}_{\mathbb{R}^n}(0,1)$ with $Dg(x)u_x < -\frac{\lambda}{2}$. Taylor's theorem thus yields

$$g(x+tu_x) = g(x) + tDg(x)u_x + o(t) < g(x) - \frac{\lambda}{4}t$$

for small t.

This means that we have some $x' = x + tu_x$ with

$$g(x')^+ - g(x) < -\frac{\lambda}{4}\|x-x'\|,$$

which shows calmness by Theorem 24. □

Note. The reverse of Corollary 38 is not true, i.e. the sublevel set map S may be calm at $(0, \bar{x}) \in \operatorname{gph} S$ for $g \in C^1(X, \mathbb{R})$ with $g(\bar{x}) = 0$ but $Dg(\bar{x}) = 0$ – just consider $g \equiv 0$.

Having Theorems 24 and 25 we now can show a sufficient property for calmness [q] resp. global error bounds with Hölder exponent q for sets defined by finite lower semicontinuous inequality systems. In the case of only one l.s.c. function the condition is a little weaker than for proper systems.

The result for one l.s.c. function was as well given by Wu and Ye in [WY02a, Theorem 6][12] (using a different proof) and a similar proposition was shown before by Ng and Zheng [NZ00, Theorem 3] for the sublevel set of a single weakly l.s.c. function (i.e. l.s.c. with respect to the weak topology).[13]

Theorem 39 (level set of one l.s.c. function). *Let X be a Banach space and $g : X \to \mathbb{R}$ lower semicontinuous with sublevel set*

---

[12]Compare also [WY02b, Theorem 3.1 and 4.1] and [WY04, Theorem 2.2] for error bounds with exponent 1.

[13]In particular our Theorem (again) gives an affirmative answer to the question [NZ00, Problem 1] adressed by Ng and Zheng after their statement: ›Is [NZ00, Theorem 3] true if the function is only assumed to be lower semicontinuous?‹.

4.5 Sufficient conditions

$S(p) := \{ x \in X \mid g(x) \leq p \}$ with $S(0) \neq \emptyset$. If exist $0 < q \leq 1$ and $\lambda > 0$ such that

$$\forall x \in X \setminus S(0) \ \exists u_x \in bd\, \mathbb{B}(0,1):$$

$$d^- g(x)(u_x) := \liminf_{\substack{t \to 0^+ \\ u' \to u_x}} \frac{g(x+tu') - g(x)}{t} \leq -\lambda g(x)^{1-q}, \quad (4.64)$$

then

$$\forall x \in X: \ dist(x, S(0)) \leq \frac{1}{\lambda q}(g(x)^+)^q.$$

Proof. Let $x \in X \setminus S(0)$ and take some $u_x \in bd\, \mathbb{B}(0,1)$ fulfilling the assumption. By definition of the lower subderivative d^-, there exists, for each fixed $c > 1$, a sequence $t_n \to 0^+$ s.t.

$$\frac{g(x+t_n u_x) - g(x)}{t_n} < -\frac{c-1}{c}\lambda g(x)^{1-q}. \quad (4.65)$$

Because for $c > 1$ it is $-\frac{c-1}{c}\lambda g(x)^{1-q} < 0$, we have particularly

$$g(x+t_n u_x) - g(x) < 0.$$

And since g is l.s.c., it holds $\liminf_{n \to \infty} g(x+t_n u_x) = g(x) > 0$, so $g(x+t_n u_x) > 0$ for sufficiently large n.

Now

$$g(x+t_n u_x)^q = g(x)^q + \frac{q}{g(x)^{1-q}}(g(x+t_n u_x) - g(x)) + O(t_n),$$

where

$$O(t_n) = \left(1 - q\left(\frac{g(x+t_n u_x)}{g(x)}\right)^{1-q}\right)g(x+t_n u_x)^q - (1-q)g(x)^q$$

$$= o(g(x+t_n u_x) - g(x)).$$

Hence $O(t_n) \leq -\frac{q}{cg(x)^{1-q}}(g(x+t_n u_x) - g(x))$ if n is sufficiently large, and thus, applying (4.65), one gets for such n that

$$g(x+t_n u_x)^q \leq g(x)^q + \frac{(c-1)\,q\,t_n}{c\,g(x)^{1-q}} \left(\frac{g(x+t_n u_x) - g(x)}{t_n} \right)$$

$$< g(x)^q - \frac{(c-1)^2}{c^2}\,\lambda\,q\,t_n.$$

So, for $t' = t_n > 0$ with n large enough and $x' = x + t'u_x$, it holds

$$\|x - x'\| = t' < \frac{c^2}{(c-1)^2}\frac{1}{\lambda q}\left(g(x)^q - g(x')^q\right)$$

$$= \frac{c^2}{(c-1)^2}\frac{1}{\lambda q}\left(g(x)^q - (g(x')^+)^q\right).$$

Thus Theorem 25 yields for each $c > 1$:

$$\forall\, x \in X : \ \mathrm{dist}(x, S(0)) \leq \frac{c^2}{(c-1)^2}\frac{1}{\lambda q}(g(x)^+)^q,$$

which – because of $\frac{c^2}{(c-1)^2} \xrightarrow[c \to \infty]{} 1$ – implies

$$\forall\, x \in X : \ \mathrm{dist}(x, S(0)) \leq \frac{1}{\lambda q}(g(x)^+)^q. \qquad \square$$

The sufficient condition (4.64) is rather strong and not a necessary one:

Example 10. *For the differentiable function*

$$g(x) = \begin{cases} x^2 \sin(\frac{1}{x}) + x, & x \neq 0 \\ 0, & x = 0 \end{cases}$$

it is[14] $g(x) < \frac{x}{2} < 0$ *if* $x < 0$ *and* $g(x) > \frac{x}{2} > 0$ *if* $x > 0$. *So in particular* $S(0) = \{\,x \mid g(x) \leq 0\,\} = (-\infty, 0]$ *and altogether we*

---
[14]Consider the cases $0 < x < 1/\pi$ and $x \geq 1/\pi$, and use $g(-x) = -g(x)$ to see this.

4.5 Sufficient conditions

have
$$\forall\, x \notin S(0): \ dist(x, S(0)) = x < 2g(x).$$

But even locally condition (4.64) does not hold, since the derivative g' (which is not continuous in $x = 0$) is given by

$$g'(x) = \begin{cases} 2x\sin(\tfrac{1}{x}) - \cos(\tfrac{1}{x}) + 1, & x \neq 0 \\ 0, & x = 0, \end{cases}$$

and $2x\sin(\tfrac{1}{x}) - \cos(\tfrac{1}{x}) + 1$ is zero for all $x = x_k = \tfrac{1}{2k\pi}$, $k \in \mathbb{N}$.

As a generalization for solution sets of finitely many lower semicontinuous inequations we get the following theorem (for a similar proposition about closed convex subsets and convex l.s.c. functions see [WY02b, Corollary 4.2]):

Theorem 40 (finite system of l.s.c. inequalities). *Again let X be a Banach space and $S(p) := \{\, x \in X \mid \bigwedge_{i=1}^{m} g_i(x) \leq p_i \,\}$ with $p = (p_1, \ldots, p_m) \in \mathbb{R}^m$, $S(0) \neq \emptyset$ and $g_i : X \to \mathbb{R}$, $i = 1, \ldots, m$, lower semicontinuous functions.*

Further let $0 < q \leq 1$ and $\lambda > 0$ be given such that

$$\forall\, x \in X \backslash S(0) \ \exists\, u_x \in bd\, \mathbb{B}(0,1) \ \forall\, i \in I(x):$$
$$g_i'(x; u_x) := \lim_{t \to 0^+} \frac{g_i(x + tu_x) - g_i(x)}{t} \leq -\lambda g_i(x)^{1-q}; \qquad (4.66)$$

where

$$I(x) := \{\, i \in \{1, \ldots, m\} \mid g_i(x) = \max_i g_i(x)^+ \,\}.$$

Then

$$\forall\, x \in X: \ dist(x, S(0)) \leq \frac{1}{\lambda q}\bigl(\max_{i=1,\ldots,n} g(x)^+\bigr)^q.$$

4 Hölder calmness – conditions and characterizations

Proof. Put $f(x) := \max_{i=1,\ldots,m} g_i(x)^+$ which is (as max-function of l.s.c. functions) again l.s.c. As in the proof of Theorem 39 above one gets for every $c > 1$ some sufficiently small $t' > 0$ s.t. for each $i \in I(x)$

$$g_i(x + t'u_x)^q < g_i(x)^q - \frac{(c-1)^2}{c^2}\lambda q\, t' = f(x)^q - \frac{(c-1)^2}{c^2}\lambda q\, t'.$$

Here we need the (one-sided) directional derivative $g_i'(x; u_x) \le -\lambda g_i(x)^{1-q}$ instead of $d^-g_i(x)(u_x) \le -\lambda g_i(x)^{1-q}$ since the latter would only give this inequation for (may be different[15]) $t_i' > 0$ and we could not use the following argument:

By definition of f and $I(x)$, and since $\lim_{t\to 0^+} g_i(x + tu_x) = g_i(x) = f(x)$ for $i \in I(x)$, it is $g_i(x + t'u_x) = f(x + t'u_x)$ for at least one $i \in I(x)$, so the above inequality implies

$$\|x - x'\| < \frac{c^2}{(c-1)^2}\frac{1}{\lambda q}\left(f(x)^q - (f(x')^+)^q\right)$$

for $x' = x + t'u_x$ and thus by Theorem 25 (and the same argument as in the proof of Theorem 39)

$$\forall\, x \in X : \operatorname{dist}(x, S_f(0)) \le \frac{1}{\lambda q}\left(f(x)^+\right)^q,$$

where $S_f(r) := \{\, x \mid f(x) \le r \,\}$. Now $S_f(0) = S(0)$ yields the proposition. □

Next we will modify conditions (4.64) and (4.66) in such a way that we will use the distance of points x outside of $S(0)$ to some (known) point $\bar{x}$ in this set instead of the function value of x. These conditions are of course less direct to verify but will prove helpful in the following.

---

[15] Consider crisscrossing zigzag functions g_i.

4.5 Sufficient conditions

Theorem 41. *Having the same setting as in Theorems 39 and 40 we may replace (4.64) and (4.66) by*

$$\forall\, x \in X \backslash S(0)\ \exists\, \bar{x} \in S(0)\ \exists\, u_x \in bd\,\mathbb{B}(0,1):$$
$$d^- g(x)(u_x) \le -\lambda \|x - \bar{x}\|^{\frac{1-q}{q}}, \qquad (4.67)$$

resp.

$$\forall\, x \in X \backslash S(0)\ \exists\, \bar{x} \in S(0)\ \exists\, u_x \in bd\,\mathbb{B}(0,1)\ \forall\, i \in I(x):$$
$$g_i'(x; u_x) \le -\lambda \|x - \bar{x}\|^{\frac{1-q}{q}}, \qquad (4.68)$$

as sufficient conditions for

$$\forall\, x \in X:\ dist(x, S(0)) \le \max\{\tfrac{1}{\lambda q}, 1\}(g(x)^+)^q$$

and

$$\forall\, x \in X:\ dist(x, S(0)) \le \max\{\tfrac{1}{\lambda q}, 1\}\big(\max_{i=1,\ldots,n} g(x)^+\big)^q,$$

respectively.

Proof. We will only give the proof for the first part of the theorem, the second one follows in the same way (cf. the proof of Theorem 40).

Consider $x \in X \setminus S(0)$ and let $\bar{x}$ and u_x as in (4.67). If $\|x - \bar{x}\| < g(x)^q$ put $x' = \bar{x}$, so

$$\big(g(x')^+\big)^q = 0 < g(x)^q - \|x - \bar{x}\|.$$

Suppose $\|x - \bar{x}\| \ge g(x)^q$ in the following.

As in the proof of Theorem 39 one gets $g(x + t' u_x) > 0$,

$$\frac{g(x + t' u_x) - g(x)}{t'} < -\frac{c-1}{c}\lambda \|x - \bar{x}\|^{\frac{1-q}{q}}$$

4 Hölder calmness – conditions and characterizations

and

$$g(x+t'u_x)^q \leq g(x)^q + \frac{c-1}{c}\frac{q\,t'}{g(x)^{1-q}}\left(\frac{g(x+t'u_x)-g(x)}{t'}\right)$$

for any $c > 1$ and some $t' > 0$ small enough. These properties yield

$$g(x+t'u_x)^q < g(x)^q - \frac{(c-1)^2}{c^2}\frac{\lambda\,q\,t'}{g(x)^{1-q}}\|x-\bar{x}\|^{\frac{1-q}{q}}.$$

Since by assumption $-\frac{\|x-\bar{x}\|^{\frac{1}{q}}}{g(x)} \leq -1$, it is also $-\left(\frac{\|x-\bar{x}\|^{\frac{1}{q}}}{g(x)}\right)^{1-q} \leq -1$, and thus it follows

$$g(x+t'u_x)^q < g(x)^q - \frac{(c-1)^2}{c^2}\lambda\,q\,t'$$

and with $x' = x + t'u_x$ we have

$$\left(g(x')^+\right)^q < g(x)^q - \frac{(c-1)^2}{c^2}\lambda\,q\,\|x-x'\|.$$

In any case, with $L_c = \max\{\frac{(c-1)^2}{c^2}\frac{1}{\lambda q},1\}$, we find some $x' \in X$ such that $\|x-x'\| < L_c\left(g(x)^q - \left(g(x')^+\right)^q\right)$ and thus Theorem 25 and a limiting process yield the desired proposition. □

As a immediate corollary we get (local) Hölder calmness:

Corollary 42. *Let X be a Banach space and $g_i : X \to \mathbb{R}$, $i = 1,\ldots,m$, lower semicontinuous functions.*
Then $S(p) := \{\,x \in X \mid \wedge_{i=1}^m g_i(x) \leq p_i\,\}$ is calm $[q]$ with rank $\frac{1}{\lambda q}$ at $(0,\bar{x}) \in \mathrm{gph}\,S$ if – with $I(x)$ as in Theorem 40 –

$$\exists\,\varepsilon,\lambda > 0\;\forall\,x \in \mathbb{B}(\bar{x},\varepsilon)\setminus S(0)\;\exists\,u_x \in \mathrm{bd}\,\mathbb{B}(0,1)$$
$$\forall\,i \in I(x):\; g'_i(x;u_x) \leq -\lambda\|x-\bar{x}\|^{\frac{1-q}{q}}. \tag{4.69}$$

4.5 Sufficient conditions

Note. As above we may replace the directional derivative in (4.69) with the lower subderivative d^- for the case $m = 1$.

Proof. Again we set $f(x) := \max_{i=1,\ldots,m} g_i(x)^+$. By Remark 18 it suffices to consider sequences $x_k \to \bar{x}$ with $x_k \neq \bar{x}$, $0 < f(x_k)$ and $\lim_{k \to \infty} f(x_k)^q \|x_k - \bar{x}\|^{-1} = 0$ only. In particular then $f(x_k)^q < \|x_k - \bar{x}\|$ for k large enough and $g_i(x_k) > 0$ for all $i \in I(x_k)$.

The proposition then follows from Theorem 24 using the same arguments as in Theorem 41. □

The next statement is an immediate corollary of the above statement and so does not need to be mentioned apart. But since we will use this special form later on and the proof given has its own right, we do so anyhow.

Corollary 43. *Let X be a subspace of $\mathbb{R}^n$ and*

$$S(p) := \{\, x \in X \mid \bigwedge_{i=1}^m g_i(x) \leq p_i \,\}$$

with $p = (p_1, \ldots, p_m) \in \mathbb{R}^m$, $g_i \in C^1(X, \mathbb{R})$, $i = 1, \ldots, m$. Further let $\bar{x} \in S(0)$ and $I = I(\bar{x}) := \{\, i \mid g_i(\bar{x}) = 0 \,\}$.
Then S is calm $[1/d]$ with rank $\frac{d}{\lambda}$ on X at $(0, \bar{x})$ if

$$\exists\, \varepsilon, \lambda > 0 \, \forall\, x \in \mathbb{B}_X(\bar{x}, \varepsilon) \setminus S(0) \, \exists\, u_x \in bd\, \mathbb{B}(0, 1) \\ \forall\, i \in I(x) : Dg_i(x)u_x \leq -\lambda \|x - \bar{x}\|^{d-1}, \quad (4.70)$$

where $I(x) = \{\, i \in I \mid \sqrt[d]{g_i(x)^+} = \max_{i \in I} \sqrt[d]{g_i(x)^+} \,\}$.

Note. Of course we suppose w.lo.g. $I \neq \emptyset$, since else we may choose $\varepsilon > 0$ small enough s.t. $g_i(x) < 0$ for each $i = 1, \ldots, m$ and all $x \in \mathbb{B}(\bar{x}, \varepsilon)$.

Proof. This proof uses some of the ideas presented in [Kum09, Theorem 4.11].

4 Hölder calmness – conditions and characterizations

Put $f(x) := \max_{i \in I} \sqrt[d]{g_i(x)^+}$. We will show (proper) calmness of the sublevel set map of this function which is equivalent to calmness $[1/d]$ of S (cf. Remark 16).

By Remark 18 it suffices to consider sequences $x_k \to \bar{x}$ with $x_k \neq \bar{x}$, $0 < f(x_k)$ and $\lim_{k \to \infty} f(x_k) \|x_k - \bar{x}\|^{-1} = 0$ only. In particular then $f(x_k) < \|x_k - \bar{x}\|$ for k large enough and $g_i(x_k) > 0$ for all $i \in I(x_k)$.

Now, setting $f_i(x) := \sqrt[d]{g_i(x)^+}$, it is

$$Df_i(x) = \frac{1}{d \cdot g_i(x)^{1-\frac{1}{d}}} Dg_i(x) = \frac{1}{d \cdot f_i(x)^{d-1}} Dg_i(x),$$

if $g_i(x) > 0$, and so (using Taylor's theorem)

$$f_i(x + tu) = f_i(x) + \frac{t}{d \cdot f_i(x)^{d-1}} Dg_i(x)u + o_i(t)$$
$$= f(x) + \frac{t}{d \cdot f(x)^{d-1}} Dg_i(x)u + o_i(t),$$

whenever $i \in I(x)$.

Take any $c > 1$. Then property (4.70) yields for for each $i \in I(x_k)$, that, for large k – in particular we choose k large enough s.t. $x_k \in \mathbb{B}(\bar{x}, \varepsilon)$ – and $t > 0$ small enough, holds

$$f_i(x_k + tu_{x_k}) \leq f(x_k) + \frac{c-1}{c} \frac{t}{d \cdot f(x_k)^{d-1}} Dg_i(x_k)u_{x_k}$$
$$< f(x_k) - \frac{(c-1)^2}{c^2} \frac{t}{d \cdot f(x_k)^{d-1}} \lambda \|x_k - \bar{x}\|^{d-1}$$
$$< f(x_k) - \frac{(c-1)^2}{c^2} \frac{\lambda}{d} t$$

Fix such $t' > 0$. Since $f_i(x_k + t'u_{x_k}) = f(x_k + t'u_{x_k})$ for at least one $i \in I(x_k)$ – by definition of f and $I(x_k)$ and continuity of f_i –

4.5 Sufficient conditions

this implies

$$f(x_k + t'u_{x_k}) < f(x_k) - \frac{(c-1)^2}{c^2}\frac{\lambda}{d}t',$$

which yields by Theorem 24 calmness with rank $L_c := \frac{c^2}{(c-1)^2}\frac{d}{\lambda}$ of the sublevel set map of f. A limiting process then gives rank $\frac{d}{\lambda}$ as well.

And, as $f(x) = \sqrt[d]{\max_{i \in I} g_i(x)^+}$, we get calmness [1/d] of the sublevel set of the function $\max_{i \in I} g_i(x)^+$ (cf. Remark 16) and thus Corollary 21 shows calmness [1/d] of S, whereby the rank is not changed. □

Remark 34. Condition (4.70) is in general weaker than MFCQ since $Dg_i(\bar{x})u_x$ may be zero. But for the case of proper calmness (i.e. $d = 1$) it does not say more than that MFCQ yields calmness.

Remark 35. For S defined by a single $g \in C^1(X, \mathbb{R})$ (4.70) reduces to

$$\exists \varepsilon, \lambda > 0 \, \forall x \in \mathbb{B}_X(\bar{x}, \varepsilon) \backslash S(0) \\ \exists u_x \in \operatorname{bd} \mathbb{B}(0,1) : Dg(x)u_x \leq -\lambda \|x - \bar{x}\|^{d-1}. \quad (4.71)$$

Before presenting a sufficient property for (4.71) (see Lemma 45) we want to prove another common proposition by means of Corollary 43.

Corollary 44. *Let* $g \in C^d(\mathbb{R}, \mathbb{R})$ *and* $\bar{x} \in \mathbb{R}$ *with* $g(\bar{x}) \leq 0$, $g^{(k)}(\bar{x}) = 0$ *for all* $1 \leq k < d$ *and* $g^{(d)}(\bar{x}) \neq 0$. *Then* $S(p) := \{x \in \mathbb{R} \mid g(x) \leq p\}$ *is calm* [1/d] *at* $(0, \bar{x})$.

Note. The assumption $g(\bar{x}) \leq 0$ is not important for the proof but for the definition of calmness [q] only, since we want to have $\bar{x} \in S(0)$. The same is true for the following two propositions as well.

4 Hölder calmness – conditions and characterizations

Proof. As $g^{(d)}$ is continuous and $g^{(d)}(\bar{x}) \neq 0$ there is some $\varepsilon > 0$ such that $g^{(d)}(x) \neq 0$ for all $x \in \mathbb{B}(\bar{x}, \varepsilon)$. And – using Taylor's theorem – the other preconditions yield that for some ξ_x between x and $\bar{x}$ holds

$$g'(x) = \sum_{k=0}^{d-2} \frac{(g')^{(k)}(\bar{x})}{k!}(x-\bar{x})^k + \frac{(g')^{(d-1)}(\xi_x)}{(d-1)!}(x-\bar{x})^{d-1}$$

$$= \frac{(x-\bar{x})^{d-1}}{(d-1)!} g^{(d)}(\xi_x).$$

As a result we have $|g'(x)| = \frac{|g^{(d)}(\xi_x)|}{(d-1)!}|x-\bar{x}|^{d-1}$, i.e.

$$|g'(x)| \geq \lambda |x-\bar{x}|^{d-1}$$

for all $x \in \mathbb{B}(\bar{x}, \varepsilon)$ with $\lambda = \frac{1}{(d-1)!} \min_{x \in \mathbb{B}(\bar{x}, \varepsilon)} |g^{(d)}(x)| > 0$.

Thus (for $u_x = 1$ or -1) it is $g'(x) u_x \leq -\lambda |x-\bar{x}|^{d-1}$, whereby follows calmness $[1/d]$ of the sublevel set map at $(0, \bar{x})$. □

Notation. If $g \in C^d(\mathbb{R}^n, \mathbb{R})$ then the order of partial differentiations does not play a role, so in such case we will write any k-th partial derivative, $0 \leq k \leq d$, in the form

$$\partial^\alpha g := \partial_1^{\alpha_1} \cdots \partial_n^{\alpha_n} g$$

with adequate multiindex $\alpha \in \mathbb{N}_0^n$ s.t. $|\alpha| = \sum_{i=1}^n \alpha_i = k$.

Lemma 45. *Let X be a subspace of $\mathbb{R}^n$, $g \in C^d(X, \mathbb{R})$ and $\bar{x} \in X$ with $g(\bar{x}) \leq 0$ such that $\partial^\alpha g(\bar{x}) = 0$ for all $\alpha \in \mathbb{N}_0^n$ with $1 \leq |\alpha| < d$. Further let exist some $\varepsilon > 0$ such that for all $x^1, \ldots, x^n \in \mathbb{B}_X(\bar{x}, \varepsilon)$*

$$\max_{i=1,\ldots,n} \left| \sum_{|\alpha|=d-1} \frac{1}{\alpha!} \partial^\alpha \partial_i g(x^i) u^\alpha \right| > 0 \text{ for every } u \in X \setminus \{0\}. \quad (4.72)$$

4.5 Sufficient conditions

Then (4.71) holds true and thus $S(p) := \{\, x \in X \mid g(x) \leq p \,\}$ *is calm* $[1/d]$ *on* X *at* $(0, \bar{x})$.

Proof. Since $\left|\sum_{|\alpha|=d-1} \frac{1}{\alpha!} \partial^\alpha \partial_i g(\cdot) u^\alpha\right|$ is continuous for every fixed u and $\mathbb{B}_X(\bar{x}, \varepsilon)$ is compact, there are $\tilde{x}^{i,u} \in \mathbb{B}_X(\bar{x}, \varepsilon)$ conforming $\min_{x \in \mathbb{B}_X(\bar{x}, \varepsilon)} \left|\sum_{|\alpha|=d-1} \frac{1}{\alpha!} \partial^\alpha \partial_i g(x) u^\alpha\right|$.

So by (4.72) we have for every $u \neq 0$ in X

$$0 < \max_{i=1,\ldots,n} \left| \sum_{|\alpha|=d-1} \frac{1}{\alpha!} \partial^\alpha \partial_i g(\tilde{x}^{i,u}) u^\alpha \right|$$

$$= \max_{i=1,\ldots,n} \min_{x \in \mathbb{B}_X(\bar{x},\varepsilon)} \left| \sum_{|\alpha|=d-1} \frac{1}{\alpha!} \partial^\alpha \partial_i g(x) u^\alpha \right|,$$

and so

$$\lambda := \min_{\|u\|_X=1} \max_{i=1,\ldots,n} \min_{x \in \mathbb{B}_X(\bar{x},\varepsilon)} \left| \sum_{|\alpha|=d-1} \frac{1}{\alpha!} \partial^\alpha \partial_i g(x) u^\alpha \right| > 0.$$

Due to $\partial^\alpha g(\bar{x}) = 0$ for all $\alpha \in \mathbb{N}_0^n$ with $|\alpha| < d$ Taylor's theorem yields

$$\partial_i g(x) = \sum_{|\alpha|=d-1} \frac{1}{\alpha!} \partial^\alpha \partial_i g(\xi_x^i)(x - \bar{x})^\alpha$$

for some ξ_x^i between x and $\bar{x}$. Thus

$$\max_{\|u\|=1} |Dg(x)u| \geq \max_{i=1,\ldots,n} |\partial_i g(x)|$$

$$= \max_{i=1,\ldots,n} \left| \sum_{|\alpha|=d-1} \frac{1}{\alpha!} \partial^\alpha \partial_i g(\xi_x^i) \left(\frac{x - \bar{x}}{\|x - \bar{x}\|} \right)^\alpha \right| \|x - \bar{x}\|^{d-1}$$

$$\geq \lambda \|x - \bar{x}\|^{d-1}$$

for every $x \in \mathbb{B}_X(\bar{x}, \varepsilon)$ and so (4.71) holds. □

Remark 36. $g \in C^d(X, \mathbb{R})$ such that $\partial^\alpha g(\bar{x}) = 0$ for all $\alpha \in \mathbb{N}_0^n$ with $|\alpha| < d$ is true for instance for homogeneous polynomials g of degree d at $\bar{x} = 0$. For such g moreover $\Sigma_{|\alpha|=d-1} \frac{1}{\alpha!}\partial^\alpha \partial_i g(x^i) u^\alpha = \partial_i g(u)$.

Lemma 45 (together with Corollary 38) now yields the following proposition – which is the statement of [Kum09, Theorem 4.11] for the case of only one inequality (cf. also [Kum09, Lemma 4.12]) and the conclusion of Lemma 29 above as well:

Corollary 46. *Let $g \in C^2(\mathbb{R}^n, \mathbb{R})$ such that $g(\bar{x}) \leq 0$, $Dg(\bar{x}) = 0$ and $D^2 g(\bar{x})$ regular for some $\bar{x} \in \mathbb{R}^n$.*
Then $S(p) := \{\, x \in \mathbb{R}^n \mid g(x) \leq p \,\}$ is calm $[1/2]$ at $(0, \bar{x})$.

Proof. As $D^2 g(\bar{x})$ is regular there is some $\varepsilon > 0$ such that for all $x^1, \ldots, x^n \in \mathbb{B}(\bar{x}, \varepsilon)$ the matrix

$$\begin{pmatrix} \partial_1 \partial_1 g(x^1) & \cdots & \partial_n \partial_1 g(x^1) \\ \vdots & \ddots & \vdots \\ \partial_1 \partial_n g(x^n) & \cdots & \partial_n \partial_n g(x^n) \end{pmatrix}$$

is regular as well, so in particular $\{\, (\partial_j \partial_i g(x^i))_{i=1,\ldots,n} \mid j=1,\ldots,n \,\}$ is a set of linear independent vectors. Thus (for $u \in \mathbb{R}^n$)

$$\forall i = 1, \ldots, n : \sum_{j=1}^n \partial_j \partial_i g(x^i) u_j = 0$$
$$\iff \sum_{j=1}^n u_j \left(\partial_j \partial_1 g(x^1), \ldots, \partial_j \partial_n g(x^n) \right) = (0, \ldots, 0)$$
$$\iff \forall j = 1, \ldots, n : u_j = 0$$
$$\iff u = 0.$$

This yields $\max_{i=1,\ldots,n} \left| \sum_{j=1}^n \partial_j \partial_i g(x^i) u_j \right| > 0$ for every $u \neq 0$, i.e. we have (4.72). □

Remark 37. Having Morse lemma below one may prove Corollary 46 as well using the ideas presented in Case 1 of the proof of Theorem 50.

4.5 Sufficient conditions

Proposition 47 (Morse lemma (cf. [JJT00, Theorems 2.7.2 and 2.8.2])). *Suppose $g \in C^2(\mathbb{R}^n, \mathbb{R})$ with $g(0) \leq 0$, $Dg(0) = 0$ and $D^2g(0)$ regular, and let $k \in \mathbb{N}$ be the number of negative eigenvalues (multiplicities taken into account) of $D^2g(0)$ (by regularity the number of positive eigenvalues is thus $n - k$).*

Then exist open neighbourhoods U and V of $0 \in \mathbb{R}^n$ and a C^1 diffeomorphism $\phi : U \to V$ with $\phi(0) = 0$ such that

$$f\left(\phi^{-1}(x_1, \ldots, x_n)\right) = -\sum_{i=1}^{k} x_i^2 + \sum_{i=k+1}^{n} x_i^2.$$

5 Polynomials

5.1 Level sets of polynomials

Having the above tools we want to answer the question whether or not it holds for polynomials $g : \mathbb{R}^n \to \mathbb{R}$ of degree d that $S(p) = \{x \in \mathbb{R}^n \mid g(x) \leq p\}$ is calm $[1/d]$ at any $(0, \bar{x})$ with $g(\bar{x}) = 0$.

So let
$$g(x) = \sum_{|\alpha| \leq d} a_\alpha x^\alpha$$
be a polynomial of degree d, i.e. there exists some $\bar{\alpha} \in \mathbb{N}_0^n$ with $|\bar{\alpha}| = d$ s.t. $a_{\bar{\alpha}} \neq 0$ (see page 11 for notations). Then for any $\alpha \in M_n^{d-1} := \{\alpha \in \mathbb{N}_0^n \mid |\alpha| = d-1\}$ it is for each $i = 1, \ldots, n$ and $x \in \mathbb{R}^n$

$$\partial^\alpha \partial_i g(x) := \partial_1^{\alpha_1} \cdots \partial_n^{\alpha_n} \partial_i g(x)$$
$$= \partial_1^{\alpha_1} \cdots \partial_i^{\alpha_i+1} \cdots \partial_n^{\alpha_n} g(x) = (\alpha + i)! \, a_{\alpha+i},$$

where $\alpha + i := (\alpha_1, \ldots, \alpha_i + 1, \ldots, \alpha_n)$.

By Lemma 45 it is thus interesting to analyze

$$\max_{i=1,\ldots,n} \left| \sum_{|\alpha|=d-1} \frac{(\alpha+i)!}{\alpha!} a_{\alpha+i} u^\alpha \right|,$$

i.e. we have to check whether

$$\max_{i=1,\ldots,n} \left| \sum_{|\alpha|=d-1} (\alpha_i + 1) a_{\alpha+i} u^\alpha \right| > 0 \quad \forall \, u \neq 0,$$

which is not true in such general form. But it may be possible to

5 Polynomials

restrict ourselves to expedient subspaces X of $\mathbb{R}^n$ such that

$$\max_{i=1,\ldots,n} \left| \sum_{|\alpha|=d-1} (\alpha_i + 1) a_{\alpha+i} u^\alpha \right| > 0 \quad \forall u \in X \setminus \{0\}.$$

First let us check some easy properties of the given setting:

Remark 38. Let $d > 0$ and $g(x) = \sum_{|\alpha|=d} a_\alpha x^\alpha$.

a) Then

$$\sum_{|\alpha|=d-1} (\alpha_i + 1) a_{\alpha+i} x^\alpha = \partial_i g(x)$$

and

$$g(x) = \frac{1}{d}(x_1, \ldots, x_n) \begin{pmatrix} \sum_{|\alpha|=d-1}(\alpha_1 + 1) a_{\alpha+1} x^\alpha \\ \vdots \\ \sum_{|\alpha|=d-1}(\alpha_n + 1) a_{\alpha+n} x^\alpha \end{pmatrix}$$

with

$$\begin{pmatrix} \sum_{|\alpha|=d-1}(\alpha_1 + 1) a_{\alpha+1} x^\alpha \\ \vdots \\ \sum_{|\alpha|=d-1}(\alpha_n + 1) a_{\alpha+n} x^\alpha \end{pmatrix} = \begin{pmatrix} (\alpha_1^1 + 1) a_{\alpha^1+1} & \cdots & (\alpha_1^m + 1) a_{\alpha^m+1} \\ \vdots & \ddots & \vdots \\ (\alpha_n^1 + 1) a_{\alpha^1+n} & \cdots & (\alpha_n^m + 1) a_{\alpha^m+n} \end{pmatrix} \begin{pmatrix} x^{\alpha^1} \\ \vdots \\ x^{\alpha^m} \end{pmatrix},$$

where

$$M_n^{d-1} = \{\alpha^1, \ldots, \alpha^m\}, \quad m = \binom{n + (d-1) - 1}{d-1}.$$

Moreover for $f(x) = g(x) + \sum_{|\alpha|<d} a_\alpha x^\alpha$ one has (for arbitrary y)

$$\sum_{|\alpha|=d-1} \frac{1}{\alpha!} \partial^\alpha \partial_i f(y) x^\alpha = \partial_i g(x).$$

5.1 Level sets of polynomials

b) Obviously part a) yields

$$\max_{i=1,\ldots,n}\left|\sum_{|\alpha|=d-1}(\alpha_i+1)a_{\alpha+i}x^\alpha\right|=0$$

$$\iff \begin{pmatrix}(\alpha_1^1+1)a_{\alpha^1+1} & \cdots & (\alpha_1^m+1)a_{\alpha^m+1}\\ \vdots & \ddots & \vdots \\ (\alpha_n^1+1)a_{\alpha^1+n} & \cdots & (\alpha_n^m+1)a_{\alpha^m+n}\end{pmatrix}\begin{pmatrix}x^{\alpha^1}\\ \vdots \\ x^{\alpha^m}\end{pmatrix}=0.$$

So in particular for $g(x)=\sum_{|\alpha|=d}a_\alpha x^\alpha$ holds

$$\max_{i=1,\ldots,n}\left|\sum_{|\alpha|=d-1}(\alpha_i+1)a_{\alpha+i}x^\alpha\right|=0 \implies g(x)=0.$$

c) For $d=1$ it is $m=1$, $g(x)=a^Tx$, $a=(a_{(1,0,\ldots,0)},\ldots,a_{(0,\ldots,0,1)})^T$ and

$$\begin{pmatrix}(\alpha_1^1+1)a_{\alpha^1+1}\\ \vdots \\ (\alpha_n^1+1)a_{\alpha^1+n}\end{pmatrix}=\begin{pmatrix}a_{(1,0,\ldots,0)}\\ \vdots \\ a_{(0,\ldots,0,1)}\end{pmatrix}=Dg(0).$$

And for $d=2$ we have $m=n$ and $g(x)=x^TAx$ with symmetric matrix

$$A=\begin{pmatrix}a_{(2,0,\ldots,0,0)} & \cdots & \frac{1}{2}a_{(1,0,\ldots,0,1)}\\ \vdots & \ddots & \vdots \\ \frac{1}{2}a_{(1,0,\ldots,0,1)} & \cdots & a_{(0,0,\ldots,0,2)}\end{pmatrix}.$$

So here it is

$$\begin{pmatrix}(\alpha_1^1+1)a_{\alpha^1+1} & \cdots & (\alpha_1^n+1)a_{\alpha^n+1}\\ \vdots & \ddots & \vdots \\ (\alpha_n^1+1)a_{\alpha^1+n} & \cdots & (\alpha_n^n+1)a_{\alpha^n+n}\end{pmatrix}$$

$$=\begin{pmatrix}2a_{(2,0,\ldots,0,0)} & \cdots & a_{(1,0,\ldots,0,1)}\\ \vdots & \ddots & \vdots \\ a_{(1,0,\ldots,0,1)} & \cdots & 2a_{(0,0,\ldots,0,2)}\end{pmatrix}=D^2g(0).$$

5 Polynomials

The previous considerations yield

Lemma 48. *For homogeneous polynomials $g : \mathbb{R}^n \to \mathbb{R}$ of the form $g(x) = \sum_{i=1}^n a_i x_i^d$, with $a_i \neq 0$ for each i, it is for all $x \neq 0$*

$$\max_{i=1,\ldots,n} \left| \sum_{|\alpha|=d-1} (\alpha_i + 1) a_{\alpha+i} x^\alpha \right| = \max_{i=1,\ldots,n} \left| \partial_i g(x) \right| = \max_{i=1,\ldots,n} \left| d a_i x_i^{d-1} \right| > 0.$$

So by Lemma 45 (and Remark 36) it follows calmness $[1/d]$ for the sublevel set maps of such homogeneous polynomials.

Note. The above homogeneous polynomials have a very special shape. But we cannot go without limitations as Example 13 will show.

For at most quadratic polynomials we do not need the special form of Lemma 48 directly[1] and not even homogeneity:

Theorem 49. *If $g(x) = \sum_{|\alpha| \leq d} a_\alpha x^\alpha$ is a polynomial of degree d at most 2, then its sublevel set map $S(p) = \{ x \in \mathbb{R}^n \mid g(x) \leq p \}$ is (at least) calm $[1/2]$ at any $(0, \bar{x})$ with $g(\bar{x}) = 0$.*

Proof. Since calmness $[q]$ of sublevel set maps is stable under translation (cf. Lemma 22) and of course translation of a polynomial gives a polynomial of the same degree, we may w.l.o.g. suppose $\bar{x} = 0$. So $g(0) = 0$, i.e. $a_{(0,\ldots,0)} = 0$.

Further, as for $Dg(0) \neq 0$ we would already have calmness (cf. Corollary 38), we may assume $Dg(0) = 0$, which yields $a_\alpha = 0$ for all $\alpha \in \mathbb{N}_0^n$ with $|\alpha| = 1$.

---

[1] But as we will see in Theorem 50 below, in the quadratic case we may translate every polynomial into this shape – which we use to prove the global quadratic error bound. So implicitly we apply this special structure as well.
For a general translation see Proposition 47.

5.1 Level sets of polynomials

Thus w.l.o.g. let $g(x) = \Sigma_{|\alpha|=2} a_\alpha x^\alpha = \frac{1}{2} x^T D^2 g(0) x$. It is (cf. Remark 38)

$$\max_{i=1,\ldots,n} \left| \sum_{|\alpha|=1} (\alpha_i + 1) a_{\alpha+i} x^\alpha \right| > 0$$

iff

$$x \notin \ker D^2 g(0) := \{ x \mid D^2 g(0) x = 0 \} =: Y,$$

so by Lemma 45 the map $S_X(p) = \{ x \in X \mid g(x) \leq p \}$ is calm $[1/2]$ at $(0,0)$ on $X := Y^\perp$. And, since for all $x \in X$, $y \in Y$ it is $g(x+y) = g(x)$, Lemma 23 yields that S is calm $[1/2]$ at $(0,0)$. □

Unfortunately it is not possible to transfer this proof even for simple monomials of degree greater than 2:

Example 11. *Let $g(x) = x_1^{d-1} x_2 \in \mathbb{R}[X_1, X_2]$ with $d \geq 3$.*
Then

$$\max_{i=1,\ldots,n} \left| \sum_{|\alpha|=d-1} (\alpha_i + 1) a_{\alpha+i} x^\alpha \right| = \max\{|(d-1) x_1^{d-2} x_2|, |x_1^{d-1}|\},$$

which is 0 if and only if $x_1 = 0$. Thus the only subspaces X such that Lemma 45 is applicable are $X = \operatorname{span}\{(1,t)\}$, $t \in \mathbb{R}$, and so $X^\perp = \operatorname{span}\{(-t,1)\}$.
Thus for any $x \in X$ and $y \in X^\perp$ we have in case of

1. *$t = 0$: $g(x) = 0$ but $\operatorname{img} g = \mathbb{R}$;*

2. *d odd, $t \neq 0$: $g(y) = g(-ty_2, y_2) = (-t)^{d-1} y_2^d > 0$ for $y \in \{ y \in X^\perp \mid y_2 > 0 \}$;*

3. *d even, $t < 0$: $g(y) > 0$ for any $y \in \{ y \in X^\perp \mid y_2 \neq 0 \}$;*

4. *d even, $t > 0$: $g(1,t) = t^{d-1} > 0$ but $g((1,t) + (-2, \frac{2}{t})) = g(-1, t + \frac{2}{t}) = -(t + \frac{2}{t}) < 0$.*

In any case the conditions of Lemma 23 do not hold.

Moreover the following example – derived and generalized from [LP94, Example 4.2][2] which is concerned with quadratic systems – shows that calmness [1/d] does not hold in general for sublevel set maps of polynomials of degree d greater than 3:

Example 12.

a) Consider $g(x_1, x_2, x_3) = (x_1^k - x_2)^2 + (x_2^k - x_3)^2 + x_3^2$ with $k \in \mathbb{N}$, which is a polynomial of degree $2k$. For its sublevel set map S obviously holds $S(0) = \{0\}$, so $dist(x, S(0)) = \|x\|$ for each $x \in \mathbb{R}^3$.

Thus for $x^\varepsilon = (\varepsilon, \varepsilon^k, \varepsilon^{k^2})$, $0 < \varepsilon < 1$, one has (using maximum norm)

$$dist(x^\varepsilon, S(0)) = \varepsilon \text{ and } g(x^\varepsilon)^+ = \varepsilon^{2 \cdot k^2},$$

which yields for $k \geq 2$

$$\frac{dist(x^\varepsilon, S(0))}{\sqrt[2k]{g(x^\varepsilon)^+}} = \frac{1}{\varepsilon^{k-1}} \xrightarrow[\varepsilon \to 0]{} \infty.$$

b) For polynomials of odd degree one has to make a little more effort:

Let $g(x_1, x_2, x_3) = (x_1^k - x_2)^2 + (x_2^k - x_3)^2 + x_3^2 + x_3^{2k+1}$ with $k \in \mathbb{N}$, a polynomial of degree $2k+1$. Here we have $S(0) \cap \mathbb{B}(0,1) = \{0\}$ (since for $0 < |x_3| < 1$ it is $x_3^{2k+1} < x_3^2 \leq (x_1^k - x_2)^2 + (x_2^k - x_3)^2 + x_3^2$, i.e. $0 < g(x)$ for all $x \in \mathbb{B}(0,1) \setminus \{0\}$) and thus $dist(x, S(0)) = \|x\|$ for each $x \in \mathbb{B}(0,1)$.

So we get for $x^\varepsilon = (\varepsilon, \varepsilon^k, \varepsilon^{k^2})$, $0 < \varepsilon < 1$ (using maximum norm)

$$dist(x^\varepsilon, S(0)) = \varepsilon \text{ and } g(x^\varepsilon)^+ = \varepsilon^{2 \cdot k^2} + \varepsilon^{(2k+1) \cdot k^2} \leq 2\varepsilon^{2 \cdot k^2},$$

---
[2] cf. also [LPR96, 2.3.13 Example]

5.1 Level sets of polynomials

which yields for $k \geq 2$

$$\frac{dist(x^\varepsilon, S(0))}{\sqrt[2k+1]{g(x^\varepsilon)^+}} \geq \frac{\varepsilon^{1-\frac{2 \cdot k^2}{2k+1}}}{\sqrt[2k+1]{2}} \xrightarrow[\varepsilon \to 0]{} \infty,$$

as $2 \cdot t^2 > 2t + 1$ for $t \in \mathbb{R} \setminus [\frac{1-\sqrt{3}}{2}, \frac{1+\sqrt{3}}{2}]$.

Even more, it is not possible to obtain that mere homogeneous polynomials define in general everywhere Hölder calm sets of same exponent as their (inverse) degree:[3]

Example 13. *Take*

$$g(x_1, x_2, x_3, x_4) = (x_1^2 - x_2 x_4)^{2k} + (x_2^2 - x_3 x_4)^{2k} + (x_3 x_4)^{2k},$$

$k \in \mathbb{N}$, *which is an homogeneous polynomial of degree $4k$, for which obviously*

$$\begin{aligned} S(0) &:= \{\, x \in \mathbb{R}^4 \mid g(x) \leq 0 \,\} \\ &= S_=(0) := \{\, x \in \mathbb{R}^4 \mid g(x) = 0 \,\}. \end{aligned}$$

Because of $g(x) = 0$ iff $x_1 = 0 \wedge x_2 = 0 \wedge (x_3 = 0 \vee x_4 = 0)$, one concludes

$$S(0) = \{\, (0, 0, x_3, 0) \mid x_3 \in \mathbb{R} \,\} \cup \{\, (0, 0, 0, x_4) \mid x_4 \in \mathbb{R} \,\}.$$

But we do not have calmness $[1/(2k)]$ at $\bar{x} := (0, 0, 0, 1)$ for instance:
Since for all $x \in \mathbb{B}(\bar{x}, \frac{1}{2})$ it is

$$dist_\infty(x, S(0)) = \max\{|x_1|, |x_2|, |x_3|\},$$

---

[3]Nevertheless such proposition holds for general monomials and special types of homogeneous polynomials (cf. Lemma 12 and Lemma 48 respectively).

5 Polynomials

one has in particular for each $x^\varepsilon = (\varepsilon, \varepsilon^2, \varepsilon^4, 1)$, $0 < \varepsilon < \frac{1}{2}$,

$$dist_\infty(x^\varepsilon, S(0)) = \varepsilon.$$

But $g(x^\varepsilon)^+ = \varepsilon^{8k}$, *resulting in*

$$\frac{dist_\infty(x^\varepsilon, S(0))}{\sqrt[4k]{g(x^\varepsilon)^+}} = \frac{1}{\varepsilon} \xrightarrow[\varepsilon \to 0]{} \infty.$$

Another notable fact is that in the general case the bound depends also on the dimension n:

Example 14. *Consider* $g(x_1, \ldots, x_n) = \sum_{i=1}^{n-1}(x_i^2 - x_{i+1})^2 + x_n^2$, $n \in \mathbb{N}$, *and let* $S(p)$ *be its sublevel set map.*
Now for $x^\varepsilon = (\varepsilon, \varepsilon^2, \ldots, \varepsilon^{2^{n-1}})$, $0 < \varepsilon < 1$, *we get*

$$dist_\infty(x^\varepsilon, S(0)) = \varepsilon \quad \text{but} \quad g(x^\varepsilon)^+ = \varepsilon^{2^n}.$$

The following global variant of Theorem 49 has been proved already by Ng and Zheng [NZ00, Theorem 4]. In their paper they even gave a rigorous analysis which exponents are possible for different types of real quadratic polynomials. And in a preceding paper they examined quadratic functions with respect to error bounds with exponent 1 and offered conditions under which conditions the exponent 1 is possible, see [NZ01, Theorem 5.1].

A similar result for the solution set of a quadratic equation was presented also by Luo and Sturm [LS00, Theorem 3.1] using a different argument. Although their indirect proof may be easily adopted for sublevel sets of quadratic inequalities, it does not allow to derive the constant L and their argument is less general than the one used here.

Theorem 50. *Let* $g : \mathbb{R}^n \to \mathbb{R}$ *be a polynomial of degree 2 and denote* $S(p) = \{ x \in \mathbb{R}^n \mid g(x) \leq p \}$ *its sublevel set map whereby*

5.1 Level sets of polynomials

$S(0) \neq \emptyset$. Then exists some positive constant L s.t.

$$\forall x \in \mathbb{R}^n : \ dist(x, S(0)) \leq L \max\{g(x)^+, \sqrt{g(x)^+}\}. \tag{5.1}$$

More precisely it holds:

Let $\sum_{i=1}^{l} \lambda_i x_i^2 + \beta x_{l+1} + \gamma$ with $\beta \cdot \gamma = 0$ be the normal form of the quadratic polynomial g, where λ_i, $i \in I := \{1, \ldots, l\}$, are the nonzero-eigenvalues of its quadratic part.

1. If now g has a normal form without linear and constant part, i.e. $\beta = \gamma = 0$, then

$$\forall x \in \mathbb{R}^n : dist(x, S(0)) \leq \max\{\lambda^{-1}, 1\} \left(g(x)^+\right)^{1/2}, \tag{5.2}$$

with $\lambda = \min_{\substack{u \in \mathbb{R}^l \\ \|u\|=1}} \max_{i \in I} |\lambda_i u_i|$.

2. If the normal form of g contains a linear but no constant term (which means $\beta \neq 0$ but $\gamma = 0$), then

$$\forall x \in \mathbb{R}^n : dist(x, S(0)) \leq |\beta|^{-1} g(x)^+. \tag{5.3}$$

3. If the constant in the normal form of g is nonzero, then – with $\lambda^* := \max_{i \in I} |\lambda_i|$ and $\lambda_* := \min_{i \in I} |\lambda_i|$ – it holds

$$\forall x \in \mathbb{R}^n : dist_2(x, S(0)) \leq \frac{\sqrt{\lambda^*}}{2\lambda_* \sqrt{|\gamma|}} g(x)^+, \tag{5.4}$$

in the case of $\gamma < 0$, and

$$\forall x \in \mathbb{R}^n : dist_2(x, S(0)) \leq \max\left\{\frac{2}{\sqrt{\lambda_* \gamma}}, \frac{\sqrt{\lambda^*}}{\lambda_* \sqrt{\gamma}}\right\} g(x)^+, \tag{5.5}$$

otherwise.

5 Polynomials

Proof. Since g is a real polynomial of degree 2 we may write $g(x) = x^T A x + b^T x + c$ where $A \in \mathbb{R}^{n \times n}$ symmetric, $b \in \mathbb{R}^n$, $c \in \mathbb{R}$ (here of course $A \neq 0$, since $\deg g$ should be 2).

For real symmetric matrices A eigendecomposition yields that there is some orthogonal matrix R for which $R^T A R = D$, where D is a diagonal matrix containing the eigenvalues $\lambda_1, \ldots, \lambda_n$ of A (with multiplicities) on its diagonal s.t. $\lambda_1, \ldots, \lambda_l \neq 0$ and $\lambda_{l+1}, \ldots, \lambda_n = 0$.

For notational convenience we put

$$I := \{\, i \mid \lambda_i \neq 0 \,\} = \{1, \ldots, l\},$$

which is not an empty set as $A \neq 0$.

Using orthogonal rotation and translation we get the normal form

$$g(x) = f(w) := \sum_{i=1}^{l} \lambda_i w_i^2 + \beta w_{l+1} + \gamma$$

with $\beta \cdot \gamma = 0$ and $w = R^T x + v$ for some fixed $v \in \mathbb{R}^n$.

Now for $S_f(p) := \{\, w \in \mathbb{R}^n \mid f(w) \leq p \,\}$ it holds

$$\operatorname{dist}_2(w, S_f(0)) = \operatorname{dist}_2(x, S(0))$$

for all $x = R(w - v)$:

Let x and w be given with $x = R(w - v)$. For $w' \in \operatorname{bd} S_f(0)$ s.t. $\operatorname{dist}_2(w, S_f(0)) = \|w - w'\|_2$ and $x' = R(w' - v)$ it holds $g(x') = f(w') = 0$ and thus

$$\begin{aligned}
\operatorname{dist}_2(x, S(0)) &\leq \|x - x'\|_2 = \|R(w - w')\|_2 \\
&\leq \|R\|_2 \|w - w'\|_2 = \|w - w'\|_2 \\
&= \operatorname{dist}_2(w, S_f(0)),
\end{aligned}$$

with $\|R\|_2$ being the spectral norm of the matrix R (induced by the euclidean vector norm) – which is 1 since $R^T R = E$. In the

5.1 Level sets of polynomials

same way one proves $\operatorname{dist}_2(w, S_f(0)) \leq \operatorname{dist}_2(x, S(0))$.
Having this one sees (for any $L, q > 0$)

$$\forall\, w \in \mathbb{R}^n : \operatorname{dist}_2(w, S_f(0)) \leq L(f(w)^+)^q$$
$$\Longleftrightarrow$$
$$\forall\, x \in \mathbb{R}^n : \operatorname{dist}_2(x, S(0)) \leq L(g(x)^+)^q,$$

so we may assume w.l.o.g.

$$g(x) = \sum_{i=1}^{l} \lambda_i x_i^2 + \beta x_{l+1} + \gamma$$

with $\beta \cdot \gamma = 0$ and $\forall\, i \in I = \{1,\ldots,l\} : \lambda_i \neq 0$.
We distinguish the three cases: $\beta = \gamma = 0$, $\beta \neq 0$ and $\gamma \neq 0$.

Case 1 ($\beta = \gamma = 0$): Let

$$X := \{\, x \in \mathbb{R}^n \mid \forall\, j \notin I : x_j = 0 \,\}.$$

Then with $\lambda := \min_{\substack{u \in X \\ \|u\|=1}} \max_{i \in I} |\lambda_i u_i| > 0$ it holds for all $x \in X \setminus S_X(0)$

$$\begin{aligned}
\max_{\|u\|=1} |Dg(x)u| &\geq \max_{i=1,\ldots,n} |\partial_i g(x)| \\
&= \max_{i \in I} \left|2\lambda_i \frac{x_i}{\|x\|}\right| \|x\| \geq 2\lambda \|x\|.
\end{aligned} \quad (5.6)$$

Because $0 \in S_X(0)$ this yields[4] by Theorem 41

$$\forall\, x \in X : \operatorname{dist}(x, S_X(0)) \leq \max\{\lambda^{-1}, 1\}(g(x)^+)^{1/2}.$$

Now, since for each $x \in X$ and $y \in Y := X^\perp$ it is $g(x+y) = g(x)$, we get (5.2) by Lemma 23.

---

[4] Note that $\|x\| = \|x\|^{\frac{1-q}{q}}$ for $q = \frac{1}{2}$.

5 Polynomials

Case 2 ($\beta \neq 0$): In this case for each $x \in \mathbb{R}^n \setminus S(0)$

$$\max_{\|u\|=1} |Dg(x)u| \geq \max_{i=1,\ldots,n} |\partial_i g(x)| \\ = \max\{\max_{i \in I} |2\lambda_i x_i|, |\beta|\} \geq |\beta| > 0. \tag{5.7}$$

By Theorem 39 thus (5.3) holds.

Case 3 ($\gamma \neq 0$): For this case we need a little more effort and some notation:

First set $I_+ := \{\, i \in I \mid \lambda_i > 0 \,\}$ and $I_- := \{\, i \in I \mid \lambda_i < 0 \,\}$ and so

$$g(x) = \sum_{i \in I_+} \lambda_i x_i^2 + \sum_{i \in I_-} \lambda_i x_i^2 + \gamma.$$

Note that for $\lambda^* = \max_{i \in I} |\lambda_i|$ and $\lambda_* = \min_{i \in I} |\lambda_i|$, which are reals greater than 0, we have for all $x \in \mathbb{R}^n$

$$\frac{1}{\lambda^*} \sum_{i \in I} |\lambda_i| x_i^2 \leq \sum_{i \in I} x_i^2 \leq \frac{1}{\lambda_*} \sum_{i \in I} |\lambda_i| x_i^2, \tag{5.8}$$

that is with other words

$$-\frac{1}{\lambda^*} \sum_{i \in I_-} \lambda_i x_i^2 \leq \sum_{i \in I_-} x_i^2 \leq -\frac{1}{\lambda_*} \sum_{i \in I_-} \lambda_i x_i^2$$

and

$$\frac{1}{\lambda^*} \sum_{i \in I_+} \lambda_i x_i^2 \leq \sum_{i \in I_+} x_i^2 \leq \frac{1}{\lambda_*} \sum_{i \in I_+} \lambda_i x_i^2.$$

We subdivide this case into the parts $\gamma < 0$ and $\gamma > 0$:

a) ($\gamma < 0$) W.l.o.g. we may assume $I_+ \neq \emptyset$, since else $g(x) < 0$ for all $x \in \mathbb{R}^n$, i.e. $S(0) = \mathbb{R}^n$. So for each $x \in \mathbb{R}^n \setminus S(0)$ it holds

$$\lambda^* \sum_{i \in I_+} x_i^2 \geq \sum_{i \in I_+} \lambda_i x_i^2 > -\sum_{i \in I_-} \lambda_i x_i^2 - \gamma \geq -\gamma = |\gamma| > 0.$$

5.1 Level sets of polynomials

Thus for $u_x := -\hat{x}\left(\sum_{i \in I_+} x_i^2\right)^{-1/2}$, with

$$\hat{x}_i := \begin{cases} x_i, & \text{if } i \in I_+ \\ 0, & \text{else} \end{cases},$$

it is $\|u_x\|_2 = 1$ and

$$\begin{aligned} Dg(x)u_x &= -\frac{2}{\left(\sum_{i \in I_+} x_i^2\right)^{1/2}} \sum_{i \in I_+} \lambda_i x_i^2 \\ &\leq -2\lambda_* \left(\sum_{i \in I_+} x_i^2\right)^{1/2} \\ &< -\frac{2\lambda_* \sqrt{|\gamma|}}{\sqrt{\lambda^*}}, \end{aligned} \tag{5.9}$$

which yields (5.4) by Theorem 39.

b) ($\gamma > 0$) Let $x \in \mathbb{R}^n \setminus S(0)$. We consider the two subcases:

If $\sum_{i \in I_-} x_i^2 \leq \frac{\gamma}{4\lambda^*} \leq \frac{\gamma}{4\lambda_*}$, then choose $y \in \mathbb{R}^n$ such that $\sum_{i \in I_-} |\lambda_i| y_i^2 = \gamma$ and $\forall i \in I_+ : y_i = 0$ as well as $\forall i \notin I : y_i = x_i$.

Thus $g(y) = 0$ and so we get using (5.8)

$$\begin{aligned} \text{dist}_2(x, S(0))^2 &\leq \|x - y\|_2 = \sum_{i \in I_+} x_i^2 + \sum_{i \in I_-} (x_i - y_i)^2 \\ &\leq \sum_{i \in I_+} x_i^2 + \sum_{i \in I_-} x_i^2 + 2\left(\sum_{i \in I_-} x_i^2\right)^{\frac{1}{2}} \left(\sum_{i \in I_-} y_i^2\right)^{\frac{1}{2}} + \sum_{i \in I_-} y_i^2 \\ &\leq \sum_{i \in I_+} x_i^2 + \frac{\gamma}{4\lambda_*} + \frac{\sqrt{\gamma}}{\lambda_*}\left(\sum_{i \in I_-} |\lambda_i| y_i^2\right)^{\frac{1}{2}} + \frac{1}{\lambda_*}\sum_{i \in I_-} |\lambda_i| y_i^2 \\ &= \sum_{i \in I_+} x_i^2 + \frac{9\gamma}{4\lambda_*}. \end{aligned}$$

119

Additionally

$$g(x)^2 = \Big(\sum_{i \in I_+} \lambda_i x_i^2 + \sum_{i \in I_-} \lambda_i x_i^2 + \gamma\Big)^2$$
$$\geq \Big(\lambda_* \sum_{i \in I_+} x_i^2 - \lambda^* \sum_{i \in I_-} x_i^2 + \gamma\Big)^2$$
$$\geq \Big(\lambda_* \sum_{i \in I_+} x_i^2 - \lambda^* \frac{\gamma}{4\lambda^*} + \gamma\Big)^2$$
$$= \Big(\lambda_* \sum_{i \in I_+} x_i^2 + \frac{3}{4}\gamma\Big)^2 \geq \frac{3}{2}\gamma\lambda_* \sum_{i \in I_+} x_i^2 + \frac{9}{16}\gamma^2.$$

Hence it holds

$$\operatorname{dist}_2(x, S(0)) \leq \frac{2}{\sqrt{\lambda_* \gamma}} g(x). \qquad (5.10)$$

Next consider $\sum_{i \in I_-} x_i^2 > \frac{\gamma}{4\lambda^*}$, take $\check{x}_i := \begin{cases} x_i, & \text{if } i \in I_- \\ 0, & \text{else} \end{cases}$ and set $u_x := \check{x}\Big(\sum_{i \in I_-} x_i^2\Big)^{-1/2}$.

Then $\|u_x\|_2 = 1$ and for each $t \geq 0$ it holds

$$g(x + tu_x)$$
$$= \sum_{i \in I_+} \lambda_i x_i^2 + \sum_{i \in I_-} \lambda_i \Big(x_i + t\frac{x_i}{(\sum_{i \in I_-} x_i^2)^{-1/2}}\Big)^2 + \gamma$$
$$= g(x) - t \sum_{i \in I_-} |\lambda_i| x_i^2 \Big(\sum_{i \in I_-} x_i^2\Big)^{-1} \Big(2\Big(\sum_{i \in I_-} x_i^2\Big)^{1/2} + t\Big)$$
$$\leq g(x) - 2t \sum_{i \in I_-} |\lambda_i| x_i^2 \Big(\sum_{i \in I_-} x_i^2\Big)^{-1/2}$$
$$\leq g(x) - 2t\lambda_* \Big(\sum_{i \in I_-} x_i^2\Big)^{1/2} \leq g(x) - t\frac{\lambda_* \sqrt{\gamma}}{\sqrt{\lambda^*}}$$

5.1 Level sets of polynomials

which is less or equal 0 for $t > 0$ large enough.

Since $g(x) > 0$, the intermediate value theorem yields the existence of some $t_0 > 0$ with $g(x + t_0 u_x) = 0$ and hence $\mathrm{dist}(x, S(0)) \leq t_0$. As, by the above calculation,

$$-g(x) = g(x + t_0 u_x) - g(x) \leq -t_0 \frac{\lambda_* \sqrt{\gamma}}{\sqrt{\lambda^*}},$$

we get

$$\mathrm{dist}_2(x, S(0)) \leq \frac{\sqrt{\lambda^*}}{\lambda_* \sqrt{\gamma}} g(x). \tag{5.11}$$

Thus in any of the subcases (5.5) is true. $\square$

Remark 39. Since the algorithm of principal component analysis for quadrics allows for a detailed computation of the normal form, we can identify the value of β and γ and describe them in terms of the original polynomial. Thus we have a concrete proposition for L for general quadratic polynomials.

Remark 40. Let us reconsider Case 1 of the above proof, i.e. $g(x) = \sum_{i=1}^{l} \lambda_i x_i^2$, $\lambda_i \neq 0$ for all $i \in I$, to get a different constant L:

Again with $X := \{ x \in \mathbb{R}^n \mid x_{l+1} = \ldots = x_n = 0 \}$ and

$$\lambda := \min_{\substack{u \in X \\ \|u\|=1}} \max_{i \in I} |\lambda_i u_i| > 0,$$

we get (5.6) as well for the Euclidean norm and thus with $\lambda^* := \max_{i \in I} |\lambda_i|$ it holds

$$\max_{\|u\|=1} |Dg(x)u| \geq 2\lambda \|x\|_2 = 2\lambda \Big(\sum_{i=1}^{l} x_i^2\Big)^{\frac{1}{2}} \geq \frac{2\lambda}{\sqrt{\lambda^*}} \Big(\sum_{i=1}^{l} |\lambda_i| x_i^2\Big)^{\frac{1}{2}}$$

$$\geq \frac{2\lambda}{\sqrt{\lambda^*}} \Big(\sum_{i=1}^{l} \lambda_i x_i^2\Big)^{\frac{1}{2}} = \frac{2\lambda}{\sqrt{\lambda^*}} g(x)^{\frac{1}{2}}.$$

for each $x \in X \setminus S(0)$. And so by Theorem 39 and Lemma 23

$$\forall\, x \in \mathbb{R}^n : \operatorname{dist}_2(x, S(0)) \leq \frac{\sqrt{\lambda^*}}{\lambda} \left(g(x)^+\right)^{1/2}. \tag{5.12}$$

As a direct corollary of Theorem 50 one has akin to [LS00, Corollary 3.1] and with essentially the same proof:

Corollary 51. *Let $g_1 : \mathbb{R}^n \to \mathbb{R}$ be a quadratic polynomial, $g_2 : \mathbb{R}^n \to \mathbb{R}$ continuous and $S_i(p) = \{\, x \mid g_i(x) \leq p \,\}$, $i = 1, 2$. If $S_1(0) \subset S_2(0)$, then exists for each $\rho > 0$ some $L > 0$ with*

$$\forall\, x \in \mathbb{B}(0, \rho) : g_2(x)^+ \leq L \left(g_1(x)^+\right)^q, \tag{5.13}$$

where $q \in \{\frac{1}{2}, 1\}$ depends on the normal form of g_1.

Proof. For $x \in \mathbb{B}(0, \rho)$ consider $\bar{x} \in S_2(0)$ s.t. $\operatorname{dist}(x, S_2(0)) = \|x - \bar{x}\|$, i.e. in particular $g_2(\bar{x}) = 0$ if $x \notin S_2(0)$ and $\bar{x} = x$ elsewise.

Since the continuous function g_2 is Lipschitz on any compact set there is some $L_2 > 0$ with

$\forall\, x \in \mathbb{B}(0, \rho)$:
$$g_2(x)^+ = |g_2(x) - g_2(\bar{x})| \leq L_2 \|x - \bar{x}\| = L_2 \operatorname{dist}(x, S_2(0)).$$

Now due to Theorem 50 it holds for some $L_1 > 0$ and $q \in \{\frac{1}{2}, 1\}$

$$\forall\, x \in \mathbb{R}^n : \operatorname{dist}(x, S_1(0)) \leq L_1 \left(g_1(x)^+\right)^q,$$

and thus, because $S_1(0) \subset S_2(0)$ yields

$$\operatorname{dist}(x, S_2(0)) \leq \operatorname{dist}(x, S_1(0)),$$

we get

$$\forall\, x \in \mathbb{B}(0, \rho) : g_2(x)^+ \leq L_2 \operatorname{dist}(x, S_1(0)) \leq L_1 L_2 \left(g_1(x)^+\right)^q. \quad \square$$

5.2 Polynomial systems

Next we want to check solution sets of systems of polynomials.

As a basis of all following arguments we start adopting Theorem 50 to hyperplanes:

Lemma 52. *Let $g : \mathbb{R}^n \to \mathbb{R}$ be a quadratic polynomial and $f : \mathbb{R}^n \to \mathbb{R}$ an affine function defining the hyperplane $H = \{\, x \in \mathbb{R}^n \mid f(x) = 0 \,\}$. If the set $S_H := \{\, x \in H \mid g(x) \leq 0 \,\}$ is not empty, then exists a constant $L > 0$ such that*

$$\forall\, x \in H : \operatorname{dist}(x, S_H) \leq L \max\{g(x)^+, \sqrt{g(x)^+}\}. \tag{5.14}$$

Proof. Let Π_H be the Euclidean projector onto H, set $\tilde{g}(x) := g(\Pi_H(x))$ and $\tilde{S}(p) := \{\, x \mid \tilde{g}(x) \leq p \,\}$.

Obviously $S_H \subset \tilde{S}(0)$, so $\operatorname{dist}(x, \tilde{S}(0)) \leq \operatorname{dist}(x, S_H)$. Moreover one has for each $x \in H$ and $x' \in \tilde{S}(0)$ with $\operatorname{dist}_2(x, \tilde{S}(0)) = \|x - x'\|_2$ that

$$\operatorname{dist}_2(x, S(0)) \leq \|x - \Pi_H(x')\|_2 \leq \|x - x'\|_2 = \operatorname{dist}_2(x, \tilde{S}(0)),$$

because Π_H is a projection onto a convex set and $\Pi_H(x) = x$ here. Hence it is $\operatorname{dist}_2(x, S_H) = \operatorname{dist}_2(x, \tilde{S}(0))$ for all $x \in H$.

As Π_H is affine as Euclidean projector onto an affine space, the function $\tilde{g}$ (composed from a quadratic and an affine function) is quadratic. Thus by Theorem 50 there exists $L > 0$ such that

$$\forall\, x \in \mathbb{R}^n : \operatorname{dist}(x, \tilde{S}(0)) \leq L \max\{\tilde{g}(x)^+, \sqrt{\tilde{g}(x)^+}\}.$$

Now the above considerations – and the fact that $\tilde{g}(x) = g(x)$ for all $x \in H$ – yield that for every $x \in H$ it is

$$\begin{aligned}\operatorname{dist}_2(x, S_H) = \operatorname{dist}_2(x, \tilde{S}(0)) &\leq L \max\{\tilde{g}(x)^+, \sqrt{\tilde{g}(x)^+}\} \\ &= L \max\{g(x)^+, \sqrt{g(x)^+}\}.\end{aligned} \qquad \square$$

5 Polynomials

Remark 41. As in the above lemma we will in the following always refer to (5.1) instead to the case-by-case analysis, which would – depending on the structure of the quadratic function under consideration – yield the exponent $\frac{1}{2}$ or 1 in the respect error bounds. Nevertheless one may do as well a distinction of cases, since the proofs will work for both exponents.

In Lemma 52 the exponent then would be dependent on the structure of $g(\Pi_H(x))$.

Having the above lemma and Theorem 50 as well as Hoffman's error bound (Proposition 1) we are now able to prove several propositions dealing with an inequality system of one quadratic and finitely many affine functions, i.e. with quadratic functions over convex polyhedral sets.

The Theorems 53 and 56 below, together with their corollaries, extend and relate several propositions about this topic. Specifically we will get as corollaries some propositions of Luo, Pang and Ralph (see [LPR96, 2.3.10 Theorem], [LPR96, 2.3.12 Theorem] and – here with an additional condition – [LP94, Theorem 4.1], [LP94, Corollary 4.1]) which rely on nonnegativity of the quadratic system on a convex polytope. As well we will deduce the results of Luo and Sturm [LS00, Theorem 3.3 and Corollary 3.2] – without using their rather involved proofs.

As in the mentioned papers, we start with error bounds restricted to a bounded polytope:

Theorem 53. *Again let $g : \mathbb{R}^n \to \mathbb{R}$ be quadratic and $S(p) := \{\, x \mid g(x) \leq p \,\}$. Further let $P = \{\, x \in \mathbb{R}^n \mid \wedge_{i=1}^m f_i(x) \leq 0 \,\}$ be a convex and bounded polytope defined by finitely many affine functions f_i, $i = 1, \ldots, m$.*

Then, if $S_P := S(0) \cap P \neq \emptyset$, there is some constant $L > 0$ such that

$$\forall\, x \in P : \operatorname{dist}(x, S_P) \leq L \max\{g(x)^+, \sqrt{g(x)^+}\}. \qquad (5.15)$$

5.2 Polynomial systems

Proof. First we apply Theorem 50 and take $L_1 > 0$ such that

$$\forall\, x \in \mathbb{R}^n : \mathrm{dist}(x, S(0)) \leq L_1 \max\{g(x)^+, \sqrt{g(x)^+}\}. \tag{5.16}$$

Moreover, by Lemma 52, for each $H_i := \{\, x \in \mathbb{R}^n \mid f_i(x) = 0 \,\}$ exists some $\tilde{L}_i > 0$ with

$$\forall\, x \in H_i : \mathrm{dist}(x, S_{H_i}) \leq \tilde{L}_i \max\{g(x)^+, \sqrt{g(x)^+}\}.$$

Thus for $L_2 := \max_{i=1,\ldots,m} \tilde{L}_i$ it holds for each i that

$$\forall\, x \in H_i : \mathrm{dist}(x, S_{H_i}) \leq L_2 \max\{g(x)^+, \sqrt{g(x)^+}\}. \tag{5.17}$$

Now decompose the closed semialgebraic set $S(0)$ into finitely many semialgebraic closed connected components $T_1, \ldots, T_N$ (cf. for instance [Cos02, Theorem 2.23]) and put $J_1 := \{\, j \mid T_j \subset P \,\}$, $J_2 := \{\, j \mid T_j \not\subset P \wedge T_j \cap P \neq \emptyset \,\}$, $J_3 := \{\, j \mid T_j \cap P = \emptyset \,\}$.
Because P is compact, each T_j closed and J_3 finite, we have

$$\alpha := \frac{1}{2} \min_{j \in J_3} \mathrm{dist}(T_j, P) > 0.$$

Here we set $\alpha = +\infty$ if $J_3 = \emptyset$.

Define $S_\alpha := \{\, x \in \mathbb{R}^n \mid \mathrm{dist}(x, S(0)) < \alpha \,\}$, an open set. So $P \setminus S_\alpha$ is compact and thus exists

$$L_3 := \max_{x \in P \setminus S_\alpha} \frac{\mathrm{dist}(x, S_P)}{\mathrm{dist}(x, S(0))} \geq 1. \tag{5.18}$$

It follows

$$\forall\, x \in P \setminus S_\alpha : \mathrm{dist}(x, S_P) \leq L_3 \mathrm{dist}(x, S(0)) \\ \leq L_3 L_1 \max\{g(x)^+, \sqrt{g(x)^+}\}. \tag{5.19}$$

Next consider $x \in P \cap (S_\alpha \setminus S(0))$ and let $x' \in S(0)$ be such that $\operatorname{dist}(x, S(0)) = \|x - x'\|$.

If $x' \in P$, then it is

$$\operatorname{dist}(x, S_P) = \operatorname{dist}(x, S(0)) \leq L_1 \max\{g(x)^+, \sqrt{g(x)^+}\}. \quad (5.20)$$

Otherwise $x' \in T_j \setminus P$ for some $j \in J_2$ ($j \in J_3$ is not possible by choice of α). And since $x \in P$ and $x' \notin P$ there is some $\lambda \in (0, 1]$ s.t. $y = \lambda x + (1 - \lambda) x' \in \operatorname{bd} P$ – which means by definition of P that $y \in H_i$ for at least one i. Now, because T_j is connected, it is $H_i \cap T_j \neq \emptyset$ for at least one of those i, which we denote $i(j)$.

Putting this together it follows

$$\begin{aligned}
&\operatorname{dist}(x, S_P) \\
&\leq \|x - y\| + \operatorname{dist}(y, S_P) \leq \|x - x'\| + \operatorname{dist}(y, S_{H_{i(j)}}) \\
&= \operatorname{dist}(x, S(0)) + \operatorname{dist}(y, S_{H_{i(j)}}) \quad (5.21) \\
&\leq L_1 \max\{g(x)^+, \sqrt{g(x)^+}\} + L_2 \max\{g(x)^+, \sqrt{g(x)^+}\} \\
&\leq \max\{L_1, L_2\} \max\{g(x)^+, \sqrt{g(x)^+}\}.
\end{aligned}$$

Finally the estimates (5.19), (5.20) and (5.21) yield (5.15) for

$$L := \max\{L_1 L_3, \max\{L_1, L_2\}\}. \qquad \square$$

Corollary 54 ([LS00, Theorem 3.3]). *Let $g : \mathbb{R}^n \to \mathbb{R}$ be a quadratic function and $P \subset \mathbb{R}^n$ a finite convex and bounded polytope. If the zero set defined by $S_P := \{\, x \in P \mid g(x) = 0 \,\}$ is nonempty, then there exists a constant $L > 0$ such that*

$$\forall\, x \in P : \operatorname{dist}(x, S_P) \leq L \max\{|g(x)|, \sqrt{|g(x)|}\}. \quad (5.22)$$

Proof. Clearly $S_P = S_1 \cap S_2$ and $P = S_1 \cup S_2$ for the sets $S_1 = \{\, x \in P \mid g(x) \leq 0 \,\}$ and $S_2 = \{\, x \in P \mid -g(x) \leq 0 \,\}$.

Theorem 50 yields the existence of $L_1 > 0$ and $L_2 > 0$ such that

5.2 Polynomial systems

for all $x \in P$ it is

$$\text{dist}(x, S_1) \leq L_1 \max\{g(x)^+, \sqrt{g(x)^+}\},$$

and

$$\text{dist}(x, S_2) \leq L_2 \max\{(-g(x))^+, \sqrt{(-g(x))^+}\}.$$

Since in the case $x \in S_2 \setminus S_1$ it is $g(x)^+ = |g(x)|$ and (because the polytope P is convex) $\text{dist}(x, S_1) = \text{dist}(x, S_P)$, we thus get

$$\forall\, x \in S_2 \setminus S_1 : \text{dist}(x, S_P) \leq L_1 \max\{|g(x)|, \sqrt{|g(x)|}\},$$

and analogously

$$\forall\, x \in S_1 \setminus S_2 : \text{dist}(x, S_P) \leq L_2 \max\{|g(x)|, \sqrt{|g(x)|}\}.$$

So with $L = \max\{L_1, L_2\}$ it follows (5.22). $\square$

Corollary 55 ([LPR96, 2.3.10 Theorem]). *Let $g_i : \mathbb{R}^n \to \mathbb{R}$, $i = 1, \ldots, l$, be quadratic polynomials which are nonnegative on a convex polytope P defined by finitely many affine functions.*

If the set $S_P = \{\, x \in P \mid \bigwedge_{i=1}^{l} g_i(x) = 0 \,\}$ is not empty, then for any compact set $K \subset \mathbb{R}^n$, there exists a constant $L > 0$ such that

$$\forall\, x \in P \cap K : \text{dist}(x, S_P) \leq L \max\{r(x), \sqrt{r(x)}\}, \qquad (5.23)$$

where $r(x) := \sum_{i=1}^{l} |g_i(x)|$.

Note. W.l.o.g. we suppose $P \cap K \neq \emptyset$ (else (5.23) holds trivially).

Proof. We distinguish to cases:

Case 1 $S_P \cap K = \emptyset$: In this case it is $r(x) > 0$ everywhere on $P \cap K$. So by compactness of K there exists some $\alpha > 0$ s.t. $\max\{r(x), \sqrt{r(x)}\} \geq \alpha$ for all $x \in P \cap K$. And taking $\bar{x} \in S_P$

5 Polynomials

(which exists because S_P was assumed to be nonempty), we have

$$\operatorname{dist}(x, S_P) \leq \|x - \bar{x}\| \leq \max_{x \in P \cap K} \|x\| + \|\bar{x}\| =: C.$$

So with $L = \alpha^{-1} C$ it follows (5.23).

Case 2 $S_P \cap K \neq \emptyset$: Set $\tilde{g}(x) := \sum_{i=1}^{l} g_i(x)$, which is quadratic again and – because of the nonnegativity assumption – has the property that $\tilde{S}_P := \{\, x \in P \mid \tilde{g}(x) = 0 \,\} = S_P$.

Now let $\tilde{P}$ be a finite convex and bounded polytope containing K and apply Corollary 54 to $\tilde{g}$ and $P \cap \tilde{P}$, obtaining $L > 0$ s.t.

$$\forall\, x \in P \cap \tilde{P} : \operatorname{dist}(x, \tilde{S}_{P \cap \tilde{P}}) \leq L \max\bigl\{ |\tilde{g}(x)|, \sqrt{|\tilde{g}(x)|} \bigr\},$$

where $\tilde{S}_{P \cap \tilde{P}} := \tilde{S}_P \cap \tilde{P} = S_P \cap \tilde{P}$.

Because of

$$\operatorname{dist}(x, S_P) \leq \operatorname{dist}(x, \tilde{S}_{P \cap \tilde{P}}),$$

and $r(x) = |\tilde{g}(x)|$ for all $x \in P \cap K$ by the nonnegativity assumption, we hence get (5.23). □

Remark 42. Since in the above Theorem 53 and its corollaries we always have continuous residual functions r ($r(x)$ being $g(x)^+$, $|g(x)|$ and $\sum_{i=1}^{l} |g_i(x)|$ respectively) on a compact set K, there exists some constant $C > 0$ s.t.

$$r(x) \leq C \sqrt{r(x)}$$

for all $x \in K$ (choose $C := \max_{x \in K} \sqrt{r(x)}$).

Hence we may replace the term of the form $\max\{r(x), \sqrt{r(x)}\}$ with $\sqrt{r(x)}$ in (5.15), (5.22) and (5.23).

Next we use the above results to obtain extended propositions.

5.2 Polynomial systems

Theorem 56. *Let g be quadratic polynomial and $f = (f_1, \ldots, f_m)$ affine functions from $\mathbb{R}^n$ to $\mathbb{R}$. Further let*

$$S(p_1, p_2) = \{\, x \in \mathbb{R}^n \mid g(x) \leq p_1 \wedge f(x) \leq p_2 \,\}$$

and suppose $S(0) \neq \emptyset$.

Then for any compact set $K \subset \mathbb{R}^n$ exists a constant $L > 0$ such that

$$\forall\, x \in K: \ \mathrm{dist}(x, S(0)) \leq L\bigl(g(x)^+ + \|f(x)^+\|\bigr)^{\frac{1}{2}}. \qquad (5.24)$$

Note. The following proof is in parts orientated at the one given in [LPR96, 2.3.12 Theorem].

Proof. If $K \cap S(0) = \emptyset$, then we can argue as in Case 1 of the proof of Corollary 55. So we suppose $K \cap S(0) \neq \emptyset$ now. Since g is quadratic and f is affine on $\mathbb{R}^n$ we may write $g(x) = x^T Q x + b^T x + c$ for some symmetric matrix $Q \in \mathbb{R}^{n \times n}$, $b \in \mathbb{R}^n$, $c \in \mathbb{R}$.

Denote $P := \{\, x \in \mathbb{R}^n \mid f(x) \leq 0 \,\}$, $S_P := S(0)$ and let Π_P be the Euclidean projector onto P. Further consider a finite bounded convex polytope $P_K \supset K \cup \Pi_P(K)$, which is possible because the projection of a bounded set onto a convex one is bounded again by contraction of the projection.

By Theorem 53 (and Remark 42) exists $\tilde{L} > 0$ such that

$$\forall\, x \in P_K \cap P: \mathrm{dist}(x, S_P \cap P_K) \leq \tilde{L}\sqrt{g(x)^+}. \qquad (5.25)$$

Thus, since $\Pi_P(x) \in P_K \cap P$ if $x \in K$, and

$$\mathrm{dist}(x, S_P) \leq \mathrm{dist}(x, S_P \cap P_K),$$

it holds

$$\forall\, x \in K: \mathrm{dist}(\Pi_P(x), S_P) \leq \tilde{L}\sqrt{g(\Pi_P(x))^+}. \qquad (5.26)$$

5 Polynomials

Hence we get for all $x \in K$

$$\begin{aligned}
\mathrm{dist}(x, S(0)) &= \mathrm{dist}(x, S_P) \\
&\leq \|x - \Pi_P(x)\| + \mathrm{dist}(\Pi_P(x), S_P) \\
&\leq \|x - \Pi_P(x)\| + \tilde{L}\sqrt{g(\Pi_P(x))^+}.
\end{aligned} \qquad (5.27)$$

Because Π_P is the Euclidean projector onto P it is $\|x - \Pi_P(x)\|_2 = \mathrm{dist}_2(x, P)$, and therefore Hoffman's Lemma yields

$$\forall x \in \mathbb{R}^n : \|x - \Pi_P(x)\| \leq L_P \|f(x)^+\| \qquad (5.28)$$

for some $L_P > 0$.

So now we need to bound $g(\Pi_P(x))^+$. By Taylor expansion one obtains for $u := x - \Pi_P(x)$

$$g(\Pi_P(x)) = g(x) + (Qx + b)^T u + \frac{1}{2} u^T Q u$$

and thus (applying the Cauchy–Schwarz inequality in the last step) it holds

$$\begin{aligned}
\left(g(\Pi_P(x))^+\right)^{\frac{1}{2}} &\leq \left(g(x)^+ + |(Qx+b)^T u| + \frac{1}{2}|u^T Q u|\right)^{\frac{1}{2}} \\
&\leq \left(g(x)^+\right)^{\frac{1}{2}} + |(Qx+b)^T u|^{\frac{1}{2}} + \frac{1}{2^{\frac{1}{2}}}|u^T Q u|^{\frac{1}{2}} \qquad (5.29) \\
&\leq \left(g(x)^+\right)^{\frac{1}{2}} + \left(\|Qx\|_2^{\frac{1}{2}} + \|b\|_2^{\frac{1}{2}}\right)\|u\|_2^{\frac{1}{2}} + \frac{1}{\sqrt{2}}\|Q\|^{\frac{1}{2}}\|u\|_2,
\end{aligned}$$

where $\|Q\|$ is some matrix norm compatible with the Euclidean norm.

Setting $\rho := 2\max_{x \in K} \|x\|_2$ and

$$L_g := \max_{x \in K} \|Qx\|_2^{\frac{1}{2}} + \|b\|_2^{\frac{1}{2}} + \left(\frac{\rho}{2}\right)^{\frac{1}{2}} \|Q\|^{\frac{1}{2}},$$

5.2 Polynomial systems

this yields

$$\sqrt{g(\Pi_P(x))^+} \leq \sqrt{g(x)^+} + L_g \sqrt{\|x - \Pi_P(x)\|_2}. \tag{5.30}$$

Hence – substituting (5.28) and (5.30) into (5.27) – it holds for all $x \in K$:

$$\begin{aligned}
\text{dist}_2&(x, S(0)) \\
&\leq \|x - \Pi_P(x)\|_2 + \tilde{L}\left(\sqrt{g(x)^+} + L_g\sqrt{\|x - \Pi_P(x)\|_2}\right) \\
&\leq (C + \tilde{L}L_g)\sqrt{\|x - \Pi_P(x)\|_2} + \tilde{L}\sqrt{g(x)^+} \\
&\leq (C + \tilde{L}L_g)\sqrt{L_P}\sqrt{\|f(x)^+\|} + \tilde{L}\sqrt{g(x)^+} \\
&\leq L'\left(\sqrt{\|f(x)^+\|} + \sqrt{g(x)^+}\right),
\end{aligned} \tag{5.31}$$

where

$$C := \max_{x \in K} \sqrt{\|x - \Pi_P(x)\|} \quad \text{and} \quad L' := \max\{(C + \tilde{L}L_g)\sqrt{L_P}, \tilde{L}\}.$$

Now, taking

$$C' := \max_{x \notin S(0)} \left(\sqrt{\|f(x)^+\|} + \sqrt{g(x)^+}\right)\left(\|f(x)^+\| + g(x)^+\right)^{-\frac{1}{2}}$$

and $L := C'L'$, we have (5.24). □

Remark 43. Note that (5.29) also holds if we replace the exponent $\frac{1}{2}$ with 1. So, in the case that the exponent in (5.25) would be 1 (cf. Remark 41), we could apply the same proof to exponent 1 instead of $\frac{1}{2}$, getting 1 instead of $\frac{1}{2}$ as exponent in (5.24).

Remark 44. As Example 2 already showed, the restriction to a compact set is needed for Theorem 56.

Note that in general one can not bound $g(\Pi_P(x))^+$ from above with $c\,g(x)^+$, where $c > 0$ is any constant:

5 Polynomials

Example 15. *Suppose $g(x_1, x_2) := -x_1^2 - x_2^2 + 1$ and $f(x) = Ax$ with $A = \begin{pmatrix} 0 & 1 \\ 0 & -1 \end{pmatrix}$. Then $P = \{(x_1, 0) \mid x_1 \in \mathbb{R}\}$, $S(0) = \{(x_1, 0) \mid |x_1| \geq 1\}$ and $g(\Pi_P(x)) = -x_1^2 + 1$.*

Considering $\bar{x} = (1, 0)$ and $x_\varepsilon = (1-\varepsilon, 2\sqrt{\varepsilon})$ for some $2 > \varepsilon > 0$ one gets $g(x_\varepsilon) = -2\varepsilon - \varepsilon^2 < 0$ but $g(\Pi_P(x_\varepsilon)) = 2\varepsilon - \varepsilon^2 > 0$.

Corollary 57 ([LS00, Corollary 3.2]). *Let $g : \mathbb{R}^n \to \mathbb{R}$ be a quadratic function and*

$$P = \{x \in \mathbb{R}^n \mid f(x) \leq 0\}, \, f = (f_1, \ldots, f_m) \text{ affine},$$

a finite convex polytope. If the set $S_P := \{x \in P \mid g(x) = 0\}$ is nonempty, then for any compact set $K \subset \mathbb{R}^n$ there exists a constant $L > 0$ such that

$$\forall x \in K : dist(x, S_P) \leq L \left(|g(x)| + \|f(x)^+\|\right)^{\frac{1}{2}}. \tag{5.32}$$

Proof. Apply the proof of Theorem 56, using Corollary 54 instead of Theorem 53. □

Corollary 58 ([LPR96, 2.3.12 Theorem]). *Let $g_i : \mathbb{R}^n \to \mathbb{R}$, $i = 1, \ldots, l$, be quadratic polynomials which are nonnegative on the finite convex polytope P.*

If the set $S_P = \{x \in P \mid \wedge_{i=1}^l g_i(x) = 0\}$ is nonempty, then for any compact set $K \subset \mathbb{R}^n$, there exists $L > 0$ such that

$$\forall x \in K : dist(x, S_P) \leq L \left(\sum_{i=1}^l |g_i(x)| + \|f(x)^+\|\right)^{\frac{1}{2}}. \tag{5.33}$$

Proof. This follows from Corollary 57 for $g(x) := \sum_{i=1}^l g_i(x)$. □

However one does not have a similar proposition for general systems of even at most quadratic polynomials. Here (again) the bound does not only depend on the degree of the involved polyno-

5.2 Polynomial systems

mials but also on the dimension n (the example is a reformulation of [LPR96, 2.3.13 Example][5]):

Example 16. *Let $S(p) := \{\, x \in \mathbb{R}^n \mid \wedge_{i=1}^n g_i(x) = p_i \,\}$ with*

$$g_i(x_1, \ldots, x_n) := x_i^2 - x_{i+1}, \ i = 1, \ldots, n-1, \ \ and$$
$$g_n(x_1, \ldots, x_n) := x_n.$$

Then $S(0) = \{0\}$ and thus for $x^\varepsilon = (\varepsilon^2, \varepsilon^4, \ldots, \varepsilon^{2^n})$, $0 < \varepsilon < 1$, it holds (using maximum norm)

$$\operatorname{dist}(x^\varepsilon, S(0)) = \varepsilon \ \ but \ \ \|g(x^\varepsilon)\| = \varepsilon^{2^n}.$$

---

[5]cf. also [LP94, Example 4.2]

Bibliography

[ABR96]   Andradas, Carlos; Bröcker, Ludwig; Ruiz, Jesús M.: *Constructible Sets in Real Geometry*, volume 33 of *Ergebnisse der Mathematik und ihrer Grenzgebiete, 3. Folge – A Series of Modern Surveys in Mathematics*. Springer-Verlag, Berlin/Heidelberg/New York, 1996. ISBN 3-540-60451-0.

[Agm54]   Agmon, Shmuel: The Relaxation Method for Linear Inequalities. In: *Canadian Journal of Mathematics*, volume VI:pp. 382–392, 1954.

[Alt83]   Alt, Walter: Lipschitzian Perturbations of Infinite Optimization Problems. In: Fiacco, Anthony V., editor, *Mathematical Programming with Data Perturbations II*, volume 85 of *Lecture Notes in Pure and Applied Mathematics*, pp. 7–21. Dekker, New York, 1983. ISBN 978-0-8247-1789-6.

[BCR98]   Bochnak, Jacek; Coste, Michel; Roy, Marie-Françoise: *Real Algebraic Geometry*, volume 36 of *Ergebnisse der Mathematik und ihrer Grenzgebiete, 3. Folge – A Series of Modern Surveys in Mathematics*. Springer-Verlag, Berlin/Heidelberg/New York, 1998. ISBN 3-540-64663-9.

[BGK$^+$82]   Bank, Bernd; Guddat, Jürgen; Klatte, Diethard; Kum-

Bibliography

[blank] mer, Bernd; Tammer, Klaus: *Non-Linear Parametric Optimization*. Akademie-Verlag, Berlin, 1982.

[BGK+83] Bank, Bernd; Guddat, Jürgen; Klatte, Diethard; Kummer, Bernd; Tammer, Klaus: *Non-Linear Parametric Optimization*. Birkhäuser, Basel/Boston, 1983. ISBN 3-7643-1375-7.

[BIG03] Ben-Israel, Adi; Greville, Thomas N.E.: *Generalized Inverses: Theory and Applications*. CMS Books in Mathematics. Springer, New York, 2nd edition, 2003. ISBN 0-387-00293-6.

[BPR96] Basu, Saugata; Pollack, Richard; Roy, Marie-Françoise: On the Combinatorial and Algebraic Complexity of Quantifier Elimination. In: *Journal of the ACM*, volume 43(6):pp. 1002–1045, November 1996.

[BPR06] Basu, Saugata; Pollack, Richard; Roy, Marie-Françoise: *Algorithms in Real Algebraic Geometry*, volume 10 of *Algorithms and Computations in Mathematics*. Springer, Berlin/Heidelberg/New York, 2nd edition, 2006. ISBN 978-3-540-33098-1.

[BR90] Benedetti, Riccardo; Risler, Jean-Jacques: *Real algebraic and semi-algebraic sets*. Actualités mathématiques. Hermann, Paris, 1990. ISBN 2-7056-6144-1.

[BS00] Bonnans, Joseph Frédéric; Shapiro, Alexander: *Perturbation Analysis of Optimization Problems*. Springer Series in Operations Research. Springer, New York, 2000. ISBN 0-387-98705-3.

[Cla76] Clarke, Frank H.: A New Approach to Lagrange Multipliers. In: *Mathematics of Operations Research*, volume 1(2):pp. 165–174, May 1976.

[Cla83] Clarke, Frank H.: *Optimization and Nonsmooth Analysis*. Canadian Mathematical Society Series of Monographs & Advanced Texts. Wiley, New York, 1983. ISBN 0-471-87504-X.

[Cos00] Coste, Michel: *An introduction to semialgebraic geometry*. Quaderni dottorato di Ricerca in Matematica, Università di Pisa, Dipartimento di Matematica. Istituti Editoriali e Poligrafici Internazionali, Pisa/Roma, 2000. ISBN 88-8147-225-2.

[Cos02] Coste, Michel: An introduction to semialgebraic geometry. Online version, Institut de Recherche Mathématique de Rennes, October 2002. URL http://perso.univ-rennes1.fr/michel.coste/polyens/SAG.pdf.

[Eke74] Ekeland, Ivar: On the Variational Principle. In: *Journal of Mathematical Analysis and Applications*, volume 47(2):pp. 324–353, August 1974.

[FP03] Facchinei, Francisco; Pang, Jong-Shi: *Finite-Dimensional Variational Inequalities and Complementarity Problems, Volume I*. Springer Series in Operations Research. Springer, New York, 2003. ISBN 0-387-95580-1.

[Gfr87] Gfrerer, Helmut: Hölder continuity of solutions of perturbed optimization problems under Magasarian-Fromowitz Constraint Qualification. In: et al., Jürgen Guddat, editor, *Parametric Optimization and Related Topics*, pp. 113–124. Akademie-Verlag, Berlin, 1987.

[GVJ90] Guddat, Jürgen; Vasquez, Francisco Guerra; Jongen, Hubertus Th.: *Parametric Optimization: Singularities, Pathfollowing and Jumps*. Teubner and Wiley, Stuttgart (Teubner) and Chichester (Wiley), 1990. ISBN 3-519-02112-9 (Teubner) and 0-471-92807-0 (Wiley).

[HK06] Heerda, Jan; Kummer, Bernd: Characterization of Calmness for Banach space mappings. Preprint 2006-26, Institut für Mathematik, Humboldt-Universität zu Berlin, November 2006.

[HO05] Henrion, René; Outrata, Jiří V.: Calmness of constraint systems with applications. In: *Mathematical Programming Series B*, volume 104(2-3):pp. 437–464, 2005.

[Hof52] Hoffman, Alan J.: On Approximate Solutions of Systems of Linear Inequalities. In: *Journal of Research of the National Bureau of Standards*, volume 49(4):pp. 263–265, October 1952.

[Hör58] Hörmander, Lars: On the division of distributions by polynomials. In: *Arkiv för Matematik*, volume 3(6):pp. 555–568, December 1958.

[JJT00] Jongen, Hubertus Th.; Jonker, Peter; Twilt, Frank: *Nonlinear Optimization in Finite Dimensions – Morse Theory, Chebyshev Approximation, Transversality, Flows, Parametric Aspects*, volume 47 of *Nonconvex Optimization and Its Applications*. Kluver Academic Publishers, Dordrecht/Boston/London, 2000. ISBN 0-7923-6561-5.

[KK02a] Klatte, Diethard; Kummer, Bernd: Constrained Minima and Lipschitzian Penalties in Metric Spaces. In: *SIAM Journal on Optimization*, volume 13(2):pp. 619–633, 2002.

[KK02b] Klatte, Diethard; Kummer, Bernd: *Nonsmooth Equations in Optimization – Regularity, Calculus, Methods and Applications*, volume 60 of *Nonconvex Optimization and Its Applications*. Kluwer Academic Publishers, Dordrecht/Boston/London, 2002. ISBN 1-4020-0550-4.

[KK06] Klatte, Diethard; Kummer, Bernd: Stability of inclusions: Characterizations via suitable Lipschitz functions and algorithms. In: *Optimization*, volume 55(5 & 6):pp. 627–660, October 2006.

[KK07] Klatte, Diethard; Kummer, Bernd: Newton methods for stationary points: an elementary view of regularity conditions and solution schemes. In: *Optimization*, volume 56(4):pp. 441–462, August 2007.

[KK09] Klatte, Diethard; Kummer, Bernd: Optimization Methods and Stability of Inclusions in Banach Spaces. In: *Mathematical Programming Series B*, volume 117:pp. 305–330, July 2009.

[Kla85] Klatte, Diethard: On the stability of local and global optimal solutions in parametric problems of nonlinear programming. Seminarbericht Nr. 75:pp. 1–39, Sektion Mathematik, Humboldt-Universität zu Berlin, 1985.

[Kla94] Klatte, Diethard: On quantitative stability for non-isolated minima. In: *Control and Cybernetics*, volume 23(1 & 2):pp. 183–200, 1994.

[Kum88a] Kummer, Bernd: Newton's method for non-differentiable functions. In: et al., Jürgen Guddat, editor, *Advances in Mathematical Optimization*, volume 45 of *Mathematical Research*, pp. 114–125. Akademie-Verlag, Berlin, 1988. ISBN 3-05-500543-0.

[Kum88b] Kummer, Bernd: The inverse of a Lipschitz function in $\mathbb{R}^n$: Complete characterization by directional derivatives. Preprint no. 195, Humboldt-Universität zu Berlin, Sektion Mathematik, 1988.

[Kum08] Kummer, Bernd: Inclusions in general spaces: Hoelder stability, Solution schemes and Ekeland's principle. Preprint 2008-7, Institut für Mathematik, Humboldt-Universität zu Berlin, 2008.

[Kum09] Kummer, Bernd: Inclusions in general spaces: Hoelder stability, solution schemes and Ekeland's principle. In: *Journal of Mathematical Analysis and Applications*, volume 358(2):pp. 327–344, October 2009.

[Li97] Li, Wu: Abadie's Constraint Qualification, Metric Regularity, and Error Bounds for Differentiable Convex Inequalities. In: *SIAM Journal on Optimization*, volume 7(4):pp. 966–978, November 1997.

[LL94] Luo, Xiao-Dong; Luo, Zhi-Quan: Extension of Hoffman's error bound to polynomial systems. In: *SIAM Journal on Optimization*, volume 4(2):pp. 383–392, May 1994.

[Łoj58] Łojasiewicz, Stanisław: Division d'une distribution par une fonction analytique de variables réelles. In: *Comptes Rendus des Séances de l'Académie des Sciences*, volume 246:pp. 683–686, 1958. Séance du 3 Février 1958.

[Łoj59] Łojasiewicz, Stanisław: Sur la problème de la division. In: *Studia Mathematica*, volume 18:pp. 87–136, 1959.

Bibliography

[LP94]   Luo, Zhi-Quan; Pang, Jong-Shi: Error bounds for analytic systems and their applications. In: *Mathematical Programming*, volume 67:pp. 1–28, October 1994.

[LPR96] Luo, Zhi-Quan; Pang, Jong-Shi; Ralph, Daniel: *Mathematical Programs with Equilibrium Constraints*. Cambridge University Press, Cambridge, 1996. ISBN 0-521-57290-8.

[LS00]   Luo, Zhi-Quan; Sturm, Jos F.: Error Bounds For Quadratic Systems. In: Frenk, Hans; Roos, Kees; Terlaky, Tamás; Zhang, Shuzhong, editors, *High Performance Optimization*, volume 33 of *Applied Optimization*, pp. 383–404. Kluwer Academic Publishers, Dordrecht/Boston, 2000. ISBN 978-0-7923-6013-1.

[NZ00]   Ng, Kung Fu; Zheng, Xi Yin: Global error bounds with fractional exponents. In: *Mathematical Programming Series B*, volume 88(2):pp. 357–370, 2000.

[NZ01]   Ng, Kung Fu; Zheng, Xi Yin: Error Bounds for Lower Semicontinuous Functions in Normed Spaces. In: *SIAM Journal on Optimization*, volume 12(1):pp. 1–17, 2001.

[NZ03]   Ng, Kung Fu; Zheng, Xi Yin: Error Bounds of Constraint Quadratic Functions and Piecewise Affine Inequality Systems. In: *Journal of Optimization Theory and Applications*, volume 118(3):pp. 601–618, September 2003.

[Pan97] Pang, Jong-Shi: Error bounds in mathematical programming. In: *Mathematical Programming*, volume 79:pp. 299–332, 1997.

Bibliography

[Rob76] Robinson, Stephen M.: Regularity and Stability for Convex Multivalued Functions. In: *Mathematics of Operations Research*, volume 1(2):pp. 130–143, May 1976.

[RW98] Rockafellar, R. Tyrrell; Wets, Roger J-B.: *Variational Analysis*, volume 317 of *A Series of Comprehensive Studies in Mathematics*. Springer-Verlag, Berlin/Heidelberg/New York, 1998. ISBN 3-540-62772-3.

[Sch94] Scholtes, Stefan: Introduction to piecewise differentiable equations. Preprint No. 53/1994, Institut für Statistik und Mathematische Wirtschaftstheorie, Universität Karlsruhe, May 1994.

[Sei54] Seidenberg, Abraham: A New Decision Method for Elementary Algebra. In: *The Annals of Mathematics, Second Series*, volume 60(2):pp. 365–374, September 1954.

[Tar31] Tarski, Alfred: Sur les ensembles définissables de nombres réels I. In: *Fundamenta Mathematica*, volume 17:pp. 210–239, 1931.

[Tar48] Tarski, Alfred: A Decision Method for Elementary Algebra and Geometry. Technical report, RAND Corporation, Santa Monica, California, 1948. Prepared for publication by J.C.C. McKinsey.

[WY02a] Wu, Zili; Ye, Jane J.: On error bounds for lower semicontinuous functions. In: *Mathematical Programming Series A*, volume 92(2):pp. 301–314, 2002.

Bibliography

[WY02b] Wu, Zili; Ye, Jane J.: Sufficient Conditions for Error Bounds. In: *SIAM Journal on Optimization*, volume 12(2):pp. 421–435, 2002.

[WY04] Wu, Zili; Ye, Jane J.: First-Order and Second-Order Conditions for Error Bounds. In: *SIAM Journal on Optimization*, volume 14(3):pp. 621–645, 2004.

i want morebooks!

Buy your books fast and straightforward online - at one of world's fastest growing online book stores! Environmentally sound due to Print-on-Demand technologies.

Buy your books online at
www.get-morebooks.com

Kaufen Sie Ihre Bücher schnell und unkompliziert online – auf einer der am schnellsten wachsenden Buchhandelsplattformen weltweit! Dank Print-On-Demand umwelt- und ressourcenschonend produziert.

Bücher schneller online kaufen
www.morebooks.de

VDM Verlagsservicegesellschaft mbH
Heinrich-Böcking-Str. 6-8   Telefon: +49 681 3720 174   info@vdm-vsg.de
D - 66121 Saarbrücken   Telefax: +49 681 3720 1749   www.vdm-vsg.de

Printed by Books on Demand GmbH, Norderstedt / Germany